# RESISTING THE VIRTUAL LIFE

## The Culture and Politics of Information

RESISTING THE  VIRTUAL LIFE

The Culture and Politics of Information

Edited by James Brook and Iain A. Boal

CITY LIGHTS • SAN FRANCISCO

Cover design by Rex Ray
Book design by Nancy J. Peters
Typography by Harvest Graphics


Library of Congress Cataloging-in-Publication Data

Resisting the virtual life : the culture and politics of information /
    edited by James Brook & Iain Boal.
        p.    cm.
      ISBN 0-87286-299-2 (pbk.) : $15.95
        1. Information society.   2. Computers and civilization.
    3. Computer networks—Social aspects.      I. Brook, James, 1951–
    II. Boal, Iain.
    HM221.R47      1995
    303.48'33—dc20                                                    95-7962
                                                                     CIP

City Lights Books are available to bookstores through our primary distributor:
Subterranean Company, P.O. Box 160, 265 S. 5th St., Monroe, OR 97456.
503-847-5274. Toll-free orders 800-274-7826. FAX 503-847-6018. Our books
are also available through library jobbers and regional distributors. For personal
orders and catalogs, please write to City Lights Books, 261 Columbus Avenue,
San Francisco, CA 94133.

CITY LIGHTS BOOKS are edited by Lawrence Ferlinghetti and Nancy J. Peters
and published at the City Lights Bookstore, 261 Columbus Avenue, San Francisco,
CA 94133.

# CONTENTS

## DEGRADING WORK

## THE REPAINTING OF MODERN LIFE

# PREFACE

As computing becomes ubiquitous — with smart TVs, smart automobiles, smart homes, smart offices, smart factories, and smart weapons — the details of particular technologies, including virtual-reality technologies, are less interesting to examine than is the life that these technologies express and help form. The appeal of technologies is often ideological and symbolic, giving concrete expression to values like control, efficiency, utility, punctuality, speed, transparency, hierarchy, and power — values, in our view, too often detrimental to a more human life. It is these congealed values that we wanted to expose and combat when we first conceived *Resisting the Virtual Life*.

Our "complaint" is not against virtuality per se and is not made in the name of a "natural" life, stripped to a savage state that could exist only in fantasy; in fact, we recognize the merits of relationships at a remove whether these are conducted by letter, telephone, or e-mail. Technology can indeed act as prosthetic extension of human powers and communities. But virtual technologies are pernicious when their simulacra of relationships are deployed societywide as substitutes for face-to-face interactions, which are inherently richer than mediated interactions. Nowadays, the monosyllabic couch potato is joined by the information junkie in passive admiration of the little screen; this passivity is only refined and intensified by programmed "interactivity."

Likewise, we do not object in principle to the introduction of new machinery in the workplace. But like the Luddites of the first industrial revolution, we refuse to cede to capital the right to design and implement the sort of automation that deskills workers, extends managerial control over their work, intensifies their labor, and undermines their solidarity. Automation in the name of progress and "inevitable" technological change is primarily to the benefit of that same class that not so long ago forced people off the land and into factories, destroying whole ways of life in the process: "labor-saving" devices have not so much reduced labor as they have increased profits and refined class domination.

Resistance to a way of life that seems to represent a cryptoreligious ideal of our society — *especially* because this ideal is unobtainable, even within the consuming layer of the population — expresses itself in manifold ways: sabotage of computer systems; refunctioning of entertainment equipment for creative and political purposes; organizing for democracy-enhancing uses of technology; defending socially beneficial institutions that suffer as information technologies flourish, deforming or replacing them; and even a quiet turning of the back. . . . This is not the place for a catalog of tactics; we wish only to indicate that the battle has already been joined on numerous fronts, including terrains distant from the technical apparatus.

.

None can escape the insistent slogans and claims that the latest technological advances in computer-mediated communication will bring knowledge, pleasure, community, economic development, personal liberation — and even the salvation of "American civilization," if the Speaker of the House, Newt Gingrich, is to be believed. Whatever is missing from our lives will be rectified by ever greater access to the stores of data and new forms of entertainment that companies big and small are eager to sell.

Technological and market pressures will undoubtedly outlive the current cycle of hype and counterhype about "the information superhighway," the Internet, and compact-disc multimedia titles. Information and information technology are structural supports that business, government, and the military cannot dispense with — the flow of information will remain of paramount importance to the expansion and survival of the capitalist world system, as will the intensification and surveillance of labor that the new machines enhance.

Even when one admits that the imperatives of finance and power have certain drawbacks, one is still often persuaded of the promised benefits of the rewiring now underway. Distance learning, digital libraries, electronic voting, e-mail, video conferencing, remote banking, on-line chat, video-on-demand, home shopping, and telecommuting are among the lures for the "early adopter" of the "connected" home computer and the souped-up TV. Howard Rheingold, an enthusiast of "cyberspace" — the *imagined* space of electronic communication — describes the kind of life he associates with the technical capacities of networked computers:

> We do everything people do when people get together, but we do it with words on computer screens, leaving our bodies behind. Millions of us have already built communities where our identities commingle and interact electronically, independent of local time or location. The way a few of us live now might be the way a larger population will live, decades hence. ("A Slice of Life in My Virtual Community," *Big Dummy's Guide to the Internet* by Adam Gaffin with Jörg Heitkötter (Washington: Electronic Frontier Foundation 1994))

The wish to leave body, time, and place behind in search of electronic emulation of community does not accidentally intensify at a time when the space and time of everyday life have become so uncertain, unpleasant, and dangerous for so many — even if it is the people best insulated from risk who show the greatest fear. The litany of problems is a familiar one: people sorted into enclaves and ghettos, growing class and racial antagonisms, declining public services (including schools, libraries, and transportation), unemployment caused by automation and wandering capital, and so on. But the flight into cyberspace is motivated by some of the same fears and longings as the flight to the suburbs: it is another "white flight."

Greater forces help determine these attitudes. Since the beginnings of capitalism, masses of people have been required to demonstrate their readiness to go wherever their labor is required, to submit to whatever conditions their masters set, to disperse whenever no longer needed. Whether thrown off the land in order to work in factories or thrown out of factories when capital migrates to cheaper climes, people are continually called upon to be "flexible."

Now, if one is a member of the consuming layer, this flexibility can bring with it certain pleasures and privileges: those who can afford hardware, software, and entertainment and information services are indeed rewarded with at least the plausible illusion of "empowerment" — even if they too have to work harder in less secure conditions. But for most of the world's people, "the information age" and "the computer revolution" mean increasingly degraded lives in conditions that approximate slavery.

Recently, when Microsoft ran a series of television spots to advertise "the future," the company presented a more upbeat perspective on community and personal empowerment — on "better living through electronics," to paraphrase an old General Electric ad. In one spot, a montage that combines *National Geographic*–style "indigenous" faces with graphical user interfaces in French is given a hint of coherence by the encouraging voice of a young woman, who advises: "Listen, this stuff that we make is powerful. It makes you powerful. . . . Do something amazing. . . . We're in your corner. . . ." The voice track ends with a chorus of youthful voices singing "The world will never be the same again" to a tune that owes something — royalties, perhaps — to "We Are the World."

The seductions of an ill-defined "creativity" and virtual world travel (with references to the French colonialist "civilizing mission") are mixed with declarations that life will change forever. Of course, one has to wonder whether it isn't the target audience that is supposed to identify with the "primitive" people — and whether it's not the audience that is slated to be "civilized" by the corporate mission, as Microsoft proceeds to buy up the world's stock of images.

This is the ideological climate in which "personal" computers and CD-ROMs circulate as fetishes for worshipers of "the free market" and "the free flow of information." Ritual use of these objects serves to reinforce people's identification with the flows of capital and power that mysteriously rule their lives in parodies of natural disasters: a knowledge explosion, a crime wave, a stock-market crash, a Desert Storm. . . . The rites of work celebrated with these new machine-fetishes recall the kind of routine, calculable, scheduled work first developed in medieval monasteries: so is it any wonder that one overhears conversations about computing wherever one goes — or that religious fundamentalists, xenophobes, nationalists, racists, and fascists feel comfortable with the refurbished ideology of "the global village"?

Any hint that there might be regressive uses of communications networks would be anathema to the dissidents of industry discourse on the information superhighway. Groups such as Computer Professionals for Social Responsibility and the Electronic Frontier Foundation see no essential contradiction between social progress and corporate profits. While they are energetic defenders of on-line "privacy" and "open access" to information, their advocacy of civil rights in cyberspace is tempered by their industry affiliations; for example, they rarely challenge the assumptions that more computerization is necessarily a good thing and that "the free market" needs to be protected from "big government."

As what could be called "corporate libertarians," such organizations fear that "government intrusion" might disturb otherwise "efficient" market forces by making onerous demands on industry and granting the police the right to eavesdrop on business and private communications over the networks. While it is of course true that government bureaucracies routinely monitor people through checking and matching, for example, Internal Revenue Service and driver license records, only rare individuals — political dissidents or particularly unlucky criminals — find themselves targeted for direct electronic surveillance. On the other hand, the average person is subject to intense scrutiny by business: their work is monitored and controlled by machinery, their leisure activities are closely examined and programmed, and their purchases and financial transactions are tracked and analyzed.

Vice President Al Gore, an ideologue of the information superhighway, similarly boosts the concept of government-support-without-regulation, even if he does advocate allowing law-enforcement agencies to listen in on digital communications. He, too, promotes the idea that "business equals progress" — and that corporate domination of communication services and content is a guarantee of "free expression."

In March 1994 Gore gave a speech to the International Telecommunications Union in Buenos Aires in which he proselytized for "the free flow of information" from U.S.–based multinational corporations to the world's information consumers: "I have come here, 8,000 kilometers from my home, to ask you to help create a Global Information Infrastructure." He expands upon the benefits that a worldwide network would bring to "all members of the human family": early warning of natural catastrophes, improved health care, better education, resolution of environmental problems, economic competitiveness, the spread of democracy. Gore presents images of capitalist harmony, a self-regulating world of free exchange and increasing development that bring everyone into the modernized sphere. The GII "will make possible a global information marketplace" — with, we suspect, just enough free speech to guarantee commodity flows without jeopardizing corporate, expert, and police control over communication.

Gore's formulations turn delirious and revealing when he holds forth on the democratic potentials of computer networks, which he calls "networks of distributed intelligence" and which he claims "will spread participatory democracy":

> In the past, all computers were huge mainframes with a single processing unit, solving problems in sequence, one by one, each bit of information sent back and forth between the CPU [central processing unit] and the vast field of memory surrounding it. Now, we have massively parallel computers with hundreds — or thousands — of tiny self-contained processors distributed throughout the memory field, all interconnected, and together far more powerful and more versatile than even the most sophisticated single processor, because they each solve a tiny piece of the problem simultaneously and when all the pieces are assembled, the problem is solved. . . .
>
> In a sense, the GII will be a metaphor for democracy itself. Representative democracy does not work with an all-powerful central government, arrogating all decisions to itself. That is why communism collapsed.
>
> Instead, representative democracy relies on the assumption that the best way for a nation to make its political decisions is for each citizen — the human equivalent of the self-contained processor — to have the power to control his or her own life. . . .
>
> The GII will not only be a metaphor for a functioning democracy, it will in fact promote the functioning of democracy by greatly enhancing the participation of citizens in decision-making. And it will greatly promote the ability of nations to cooperate with each other. I see a new Athenian Age of democracy forged in the fora the GII will create.

Gore's reference to Athenian democracy is significant, but, like almost everything else in the speech, it requires translation. Athens was an imperialist slave society, with democratic rights enjoyed only by a privileged few. "Representative democracy" as Gore conceives it is more like "the representation of democracy," with the great majority of the world's people left to enjoy — at best — the mere image of participation in decisions that affect their lives. As in Athens, important decisions will continue to be made by the elite: the bourgeoisie and the politicians, bureaucrats, and experts in its employ. Machines mediating between citizen and ruling institution would in no way enhance individual freedom; instead, this scheme would further naturalize the force of law, regulation, procedure, and other codes of conduct while further depoliticizing the administration of society. The "intelligent island" of Singapore points the way. . . .

·

There *are* alternatives to the capitalist utopia of total communication, suppressed class struggle, and ever-increasing profits and control that forgets rather than resolves the central problems of our society. With the end of the Cold War and the gradual loss of illusions about a world system that benefits an ever-shrinking minority, we can more easily glimpse the outlines of these alternatives, if only in silhouette. Following the collapse of the mock-epic struggle between "World Communism" and "the Free World," "All Power to the Multinationals!" is not a deeply convincing slogan (which partly explains the rise of the less universalist ideologies of xenophobia, racism, and crime). But as Armand Mattelart observes in *Mapping World Communication: War, Progress, Culture* (Minneapolis: University of Minnesota Press 1994), we are

> decidedly light years from Marx and Engels's utopia, founded on the demolition of the parochial spirit of feudal society. In their *Manifesto of the Communist Party* they had solemnly proclaimed: "In place of the old local and national seclusion and self-sufficiency, we have intercourse in every direction, universal interdependence of nations. And as in material, so also in intellectual production. The intellectual creations of individual nations become common property. National one-sidedness and narrow-

mindedness become more and more impossible, and from the numerous national and local literatures, there arises a world literature."

The positive overcoming of national and social divisions — rather than the global flattening of traditional differences and manufacture of new differences and identities that facilitate market penetration — remains a goal worth pursuing. Much — one could say "everything" — seems to stand in the way, and the rhetoric of liberation attached to the information industry is, in our perspective, just one more obstacle to understanding and reshaping the world in more desirable directions. Opposing forces, some as yet unaware of each other, meet on the obscure terrain of information technology. Until now, critique and analysis of those forces have contributed little to the public discourse: the modest goal of *Resisting the Virtual Life* is to expand the terms of debate.

In assembling *Resisting the Virtual Life*, we did not intend to present a coherent theory of the politics of technological change; rather, we looked for a range of voices and perspectives that would address an almost unnameable object — "information age," "information superhighway," "cyberspace," "virtuality," and the proliferating variants — from a critical, democratic perspective. We hoped to use these pages to open a dialogue outside the official channels of discourse where only sanctioned "controversies" are taken seriously.

We have gathered, then, contributions from a diverse group that includes writers, scholars, activists, artists — even a software engineer. We suspect that few of our contributors would agree with one another on much except the need to question the general rush to rewire and build an infrastructure that would further an alienated life. Some of them participate (with varying degrees of eagerness) in the production of the new technologies; many are nonspecialists on the receiving end of information inundation. In the pages of *Resisting the Virtual Life* they investigate the connections between information technologies and the increasingly abstract, *virtual* existence that is their complement. Addressing the impact of the convergence of video, computer, and networked communications technologies on work, education, health, entertainment, art, and literature, the contributors explore the possibilities and strategies of resistance to the rewiring of the body psychic. Relations of power and dependence fostered by the new technologies are critically examined — from the consumer's embrace of information appliances and the worker's self-retooling to the transnational corporation's pursuit of greater profits and control.

At issue is the reorganization of life on all its levels — from the economic to the experiential, from world trade to cognition. *Resisting the Virtual Life* intends to provide correctives more profound than those generated by the cybernetic feedback mechanisms of "the marketplace of ideas," where scandalous defects are always answered by pseudocritiques that assure us that all is well, except for the inevitable bugs that the system itself will fix.

*J.B. & I.B.*
*San Francisco & Berkeley, December 1994*

THE NEW INFORMATION ENCLOSURES

*Iain A. Boal*

# A FLOW OF MONSTERS:
## Luddism and Virtual Technologies

Loud sounds the Hammer of Los, loud turn the Wheels of Enitharmon:
Her Looms vibrate with soft affections, weaving the Web of Life,
Out from the ashes of the Dead . . . .
　　　　　　　　　　— William Blake, *Milton*

A few years back, perhaps goaded beyond endurance by the bumper slogan "Kill Your Television," a small party of saboteurs assembled during the lunch hour at Sproul Plaza on the Berkeley campus. They set up a neat line of defunct TVs. The public had been invited and sledge-hammers thoughtfully provided. A spot of machine-breaking ensued. To the sound of imploding cathode-ray tubes these mild resisters were appre-hended by the police *while destroying their own, already broken, property.*

A week after the daylight wreckers smashed the TV sets, and not three hundred yards away, a routine but equally instructive event passed off without histrionics — indeed, without notice. At the back door of the university library, Department of Works operatives trashed dozens of functional computers, rendered obsolete by a newer model. The entire history of sabotage, when set against the maelstrom of capital-ism's planned destruction, would scarcely register in the scales.

That even such polite street-theater provoked arrests and the revulsion of the authorities is a testament to the lingering charge that carries over to our time from the Luddites, those English artisans who went to the gallows for the breaking — not of the means of consumption — but of productive property; in their case, the textile looms. The Luddites were skilled weavers who smashed the mechanical looms that were throwing them out of work and into starvation at the onset of the Industrial Revolution. Their resistance was a social movement that operated clandestinely under the banner of the mythical "General Ludd." On the basis of capital's stereotype of the Luddites as mindless, backward-looking wreckers, it is impossible to grasp the dynamic of resistance and accommodation or to explain the fact that the Luddites sometimes broke old frames while leaving new machinery intact. It all depended on the social relations of production and the conditions under which the power looms were to be used by the employer. For example, the Luddites targeted machinery that was being ganged up in batteries or was being automated for operation by children.

To be sure, the breaking of machinery by no means began with the outrages of 1811 against English mill-owners. Yet the weavers who smashed the stocking-frames lent their name to industrial sabotage, a form of direct action that is as old as exploitation itself. In the term "Luddite" there still resonates the *grande peur* of the possessing classes. In 1812, the British state had 24,000 troops and local militia out against the Luddites, more than had gone abroad under Wellington to fight Napoleon. In February of that year, during the parliamentary debate on the introduction of the Framework Bill, by which the government proposed to send the machine-breakers to the hanging tree, Lord Byron rose to speak in passionate, sardonic defense of the monsters who, "in the blindness of their ignorance, instead of rejoicing at these improvements in arts so beneficial to mankind, conceived themselves to be sacrificed to improvements in mechanism."

E. P. Thompson has traced that savage transition from a "moral economy" to the domination of the commodity. The Luddite weavers were swept aside in the dawn of this continuing catastrophe, in which all the world's resources — human and otherwise — are cast onto the market. In those years of revolution and counterrevolution, the discourse of political economy, free trade, and laissez-faire legitimated the wholesale enclosure of the commons, forcing freedom upon laborers rendered landless. It was Thomas More, in *Utopia*, who first bitterly portrayed the enclosures — the devouring of people and land by sheep at the beginning of agrarian capitalism in Europe. Karl Polanyi called this expropriation of communal property a revolution of the rich against the poor — a revolution that decimated the population and turned the soil into dust and the peasants into beggars. The Luddite movement was a moment and mode of resistance to what Polanyi termed "the great transformation" — the vast historical process of enclosure that strives for the commodification of land and labor. But, over time, enclosure was to take on many forms: the privatizing of communal land; the incarceration of production; the corralling of the dispossessed into ghettos, reservations, barracks, prisons, asylums, and schools; the sequestration of the airwaves by media monopolists; and so forth.

## The Technics of Modernization

When fabrication was enclosed in the new factories, this marked the realization of the old dream of Francis Bacon, the brilliant prophet of industrialism, whose technocratic fantasy wedding science and empire — *New Atlantis* was a kind of seventeenth-century R&D park — projected the transformation of the world into an open-ended and productive cornucopia.

Later seers of modernity foretold a future of boundless prosthetic enhancements to human powers, among them powers of communication — modes of amplification, replication, and extension. A cascade of inventions was developed to connect the metropoles to colonial outposts for military and commercial intelligence — submarine cable networks, land lines, and imperial wireless chains, among them. A British MP breathlessly announced to the Royal Colonial Institute in 1887: "Stronger than death-dealing warships, stronger than the might of devoted legions, stronger than wealth and genius of administration, stronger even than the unswerving justice of Queen Victoria's rule, are the scraps of paper that are borne in myriads over the seas, and the two or three slender wires that connect the scattered parts of her realm." The history of information technologies, however, as Brian Winston shows in *Misunderstanding Media*, has been one of gradual uncataclysmic development in response to certain persistent social relations: "The same authorities and institutions, the same capital, the same research effort which created today's world is trying also to create tomorrow's."

The infrastructure of empire and the new rail, road, and electro-mechanical links of industrial capitalism, together with brutal dislocations and reconnections forged by urbanization and labor migration, wrought, to be sure, profound changes in conceptions of time and space. Fin-de-siècle American critics had visions of retrieving lost community by the new-fangled "virtual" means — telegraph and telephone. Marshall McLuhan's "global village" is just a late gloss on this trope of communications-as-community. Actually, he took the idea from his Canadian mentor, Harold Innis, who in turn was drawing on Joseph Goebbels' notion of radio as a modern technical means of "retribalizing" Germany. It was the Nazi propagandistic use of the airwaves that drove Bertolt Brecht to his attempt at a radical application of radio.

### Cybernetics Redux

More recent information technics — among them holography, xerography, satellites, videotape, videophones, and fiber optics — are currently said to have reached a critical mass such that profound societal transformation is imminent. This is the second go-round for a new world under the banner of cybernetics, which emerged as a theory out of the Second World War — a twenty-four-hour, mechanized conflict coordinated by remote control — in which humans were analyzed as "factors" in the communication feedback circuits of men and machinery.

After the war, the blending of technological optimism and anti-communist Manicheism was embodied in the figure of the Hungarian emigré mathematician, John von Neumann. He designed the architecture for the general-purpose digital computer; he proposed the detonation of nuclear weapons in the Atlantic to improve the African climate; for modeling strategic engagement he developed the logic of game theory, whose premises Gregory Bateson described as "in a paranoidal direction and odious."

There was indeed a dark side to the brief American century, figured in film and popular culture. In the shadow of *Sputnik*, Hollywood conjured a bestiary of aliens, cyborgs, dinosaurs, social insects, and robots. A strain of technological pessimism turned apocalyptic after the U.S. defeat in Vietnam and the capital crisis of the 1970s. But in the last decade the computer — whose precursor, the programmable Jacquard loom, haunted the Luddites — has allowed the bourgeoisie to fall in love once again with the future. This time around, the electronic sublime predicted by the hyperbolists of the information revolution is announced under the sign of "virtuality."

The meaning of "the virtual" is obscure. What is its relation to "the real"? Is it somehow a mimicking? an approximation? a double? the antithesis of the real? Does it inhabit, so to speak, its own space, its own plane? (For a long time it has been a technical term in physics, and in art history in discussions of perspectivism.) In the same semantic field are a number of related notions — actual, counterfactual, specular, replicated, simulated, artificial — whose connections, with respect to the human sensorium and cultural reception, await their analyst.

The new virtuality, whatever its role in the construction of the real, is somehow reckoned a function of the wiring of the computer to video technologies in a global web of telecommunications. The emergence of video itself was accompanied by similar forecasts of distinctively new cultural relationships, some of which — independent production, self-timed viewing — have been significantly taken up. Yet the history of one of the battery of video techniques — the instant replay — is exemplary of the way any technological innovation gets inserted into prevailing social and institutional settings. Before videotape all telecasts were live. The only method of recording was by means of a kinescope — a film photographed directly from the television tube during the live telecast. The result was noticeably inferior in quality to the live original. In the case of videotape, whose industry use goes back to 1956 when the first, hugely expensive, Ampex video-recorder reached the market, the copy was soon technically indistinguishable from live coverage — now only a caption alerts the audience that what they are watching is a recording. One might call the result "virtually real."

The development of high-fidelity videotape predictably proved very attractive to the owners and managers of television. Telecasts could now be reproduced to an acceptably high quality and sold as a commodity. Secondly, video technology finessed the awkward but inherent unpredictability of real-time transmission. Live broadcasts always threaten to subvert the factitiously smooth *flow* of televisual reality — a flow that becomes naturalized and in a sense seamless, but at the same time rigidly segmented into programs in a way that mirrors a twenty-four-hour production line. "Bloopers" could now be domesticated, even packaged and sold as products in themselves. Thirdly, the prerecording of programs for subsequent editing before transmission represents an enormous tightening of control over the medium and its content. Indeed, video editing approaches the traditional power of the film director in the construction of events. The possibilities for review, selection, and tape-delay have made the video recorder one of the most important instruments of the gatekeepers of news and official history, and not just, or mainly, a weapon of the weak.

Although the instant replay is associated with sport, it did not begin there. Its first historic use occurred when the NBC cameras happened to be hooked up to an Ampex video recorder as Jack Ruby assassinated Lee Harvey Oswald. One hour later, almost 80 percent of American television sets were on, showing a continuous replay of the murder. It impressed not only the watching public but TV executives as well. Just a few weeks later — New Year's Day 1964 — instant replay was first used at a sports event, the Army-Navy football game.

But no Monday morning logic can account for the fact that the three sets of endlessly replayed video images that have become icons of their epoch involved death and near-fatal violence — the Oswald assassination, the *Challenger* explosion, and the assault on Rodney King. (The death of Kennedy has doubtless been replayed more than all of them, but it was not videotaped at the time, and was only later transferred from Super-8 film.) The mechanical and digital reproduction of these events distilled national nightmares. The networks saw to it that they took on another nightmare quality — recurrence. Replaying on video loop the moment of death or near-death over and over again served no functional, analytic purpose. Compulsive repetition gives the illusion of control over the repressed, yet conveniently meshes with the purposes of those who seek to saturate the audience with their products and symbols. Of course, there is the initial curiosity about "what actually happened" — and the fascination with images of "real" death, instead of the daily simulacra of carnage on film and television. In the case of the shuttle disaster, the incessant replaying of the moment of disintegration moved beyond curiosity and functional analysis to the point where the commercial reflexes of the news producers, aiming for maximum arousal around a potent symbol of chauvinistic technophilia, converged with deep collective anxieties, indeed paranoia.

## Paranoia and Virtuality

The effects of the linking of video to digital technologies is unclear. If we must reject the fantasies of the neocyberneticians, we must likewise reject the salesmen of the apocalypse, whose vision of totalitarian media is rooted in an essentially theological concept of omnipotence. Nevertheless, the case of instant replay prompts an important question. Are there emergent properties of this new constellation of digital machines and imaging techniques that suggest a *causal* relation between their kinds of virtuality and the production of paranoia?

Virtuality in other forms is not new. Nor is there anything new about the production of paranoia, given societies riven by divisions of gender, race, and class. And the best paranoids don't need machinery; they do it all in their heads. But paranoia comes in different flavors. That is, there is paranoia of and with power — and there is paranoia *against* power.

So we are not to be taken as proposing total novelty of experience in these new virtual spaces. Moreover, that Hollywood's digital technologists and Sony's virtual engineers should have brought us *dinosaurs* only seems a paradox. The cinema and the video screen — because they constitute the myth spaces of modernity — were always going to produce totemic monsters from out of deep space, or the sexual lizards, huge yet safely extinct, which body forth, from out of deep time, both the fears and wants of the audience, and the apparatus's own desire to make itself mythic.

One might posit a continuity with the experience of the peasant in the stained luminosity of the cathedral: grotesque corbel figures and brimstone. What is remarkable now, however, is the qualitative discontinuity introduced by a miniature, personalized, moving image hitched to the microchip. A video screen *breaks the distance* — in favor of an intimate one-to-one virtual relationship. The cyborgs on the virtual screen are then an allegory of the fear of social death and incorporation into the machine. Even the most skillful teenagers cannot master this program. The monsters pile on, in a parody of post-Fordist speedup. The body is fully in the virtual, but no longer in command. No surplus of skill in twitching a mouse will save the children.

So one primary candidate effect of the virtual technologies is the production of new monstrous viewers, who are incomplete, lacking, overwhelmed *inside*. Its corollary is a politics of resentment and a resentment of politics. The real virtual operation is to split open the subject and make it incomplete. Paranoia then flourishes on the cusp of a plenitude always under total threat — fertile ground for fascism, and no need to explain the looming madness by invoking either manipulators above the fray or Noam Chomsky's propaganda model of the capitalist media (and its inverse, the "information" model).

The information model is, in any case, false, because it parallels Ferdinand de Saussure's view of language, in which the two poles of the circuit of communication — sender and receiver — have equivalent power. Messages, on this account, flow between equals. This is the ideological assumption that has crippled linguistics and the field of communications studies. Moreover, the receivers of messages and images are not in a crucial sense independent of the medium. Television, for example, produces the televisual body — the couch potato. Different mediating technologies, in general, construct different subjects. The computer harnessed by the logic of administrative science, marketing, political advertising, or epidemiology constitutes a "population" quite removed from traditional notions about human groups in a nexus of community, kin, and social memory. All interiority and psychological depth are either effaced or reappear in the guise of "the irrational" and "the subjective." A population, in this "virtual" sense, is not the crowd, mob, people, folk, proletariat, nation, or citizenry of other discourses; its membership is defined by Boolean operation and segmented arbitrarily by age or income or shopping habits or blood group or zip code or some intersection of variables.

### Resistance and Critical Paranoia
Actually, a benign virtual space lies there, waiting — the human, critical imagination, where you may find out what you are *not*, and on your

own time. A paranoia against power, moreover, is recommended for resistance to the information enclosures, linked and coordinated by way of satellites, fiber optics, and the technics of silicon — though the resistance will feel, as E. P. Thompson said, like "whistling into a typhoon."

The refusal of television, truculence with smart machinery, the sabotage of genocidal apparatus — these are, to be sure, gestures of vitality, but in no way amount to a Luddite movement. In any case, for millions across the globe, there is no private phone, fax, or TV, let alone a computer or an automobile. And there never will be. This is not a question of progress or modernity delayed. There is nothing unmodern or archaic about the deserted ruins of Ireland, about the "ethnic" killing fields of Rwanda, about the eroded hillsides of Chiapas, about barrios and favelas without basic services or utilities. The pictures from Haiti reveal one of modernity's faces, as much as the footage from South Central L.A. The regnant narrative, whose keywords are "progress," "modernity," and "development," conjures its extinct dinosaurs, such as the Luddites, and its living coelacanths, namely the residual peasantry and a handful of quaint enclaves, "tribes" that constitute our contemporary ancestors. This is the metaphysic of modernity, whether in its classical capitalist form or in the form, until recently, of its Marxist sibling.

Those who criticize the deployment of certain modern technologies and yet flinch at the sobriquet "Luddite" are complicit with the logic of progress, fearful about being branded technophobe, or, finally, losers along with the peasantry and the doomed tribes. But it is a lie that direct action against the instruments of production has always been hopeless or that it somehow entails being "antitechnology," as if that were a possible position *in general*. The Elizabethan gig-mills were successfully suppressed for generations by legislation that followed agitation from below. The Japanese for a time gave up the gun. Captain Swing and the agricultural Luddites who smashed the threshing machines in the 1830s got themselves and their children a reprieve for half a century. The Luddite army of redressers had no leaders, and their machine-smashing was without violence — that is to say, they understood the radical distinction between life and property.

New conditions of work throw up new sites of resistance — production bottlenecks have always been weak spots, and now "just-in-time" inventory creates its own particular vulnerabilities. Most machines are abuser-friendly; computers, as one German engineer remarked, "do not like tea, coffee, coke or iron powder." On the side of consumption, mass boycotts of selected products terrify the personnel in sales and marketing. Unplugging and unwiring can doubtless be necessary tactics, but it all depends on context. Take the Xerox machine, for example: like those who labored courageously under Stalinist regimes to produce and distribute samizdat when all typewriters had to be registered with the police and every copying machine had an armed guard, dissidents all over the South are desperately short of means of replication. On the other hand, for critics in the Northern academy surfeited with information, there comes a time *when the Xeroxing has to stop.*

This is in no sense to adopt the functionalist philosophy of technology that posits an a priori neutrality for artifacts and technical systems — whereby machines merely have good uses and bad uses. Some things, it needs insisting, are without qualification vicious with respect to the flourishing of life: plutonium technology, for one, and, for another, if Barry Commoner is right, the entire global petrochemistry complex, which he says must be stopped in its tracks before it poisons life on earth. Fortunately, petroleum-based products, especially plastics, are mostly *replacive;* that is, they have been adopted to substitute for other materials such as ceramics, wood, and metal.

Most artifacts are not as stark as plutonium. All, nevertheless, have a *value-slope;* that is, they conduce to certain forms of life and of consciousness, and against others. Artifacts are congealed ideology. The computer, as designed, embodies the command-and-control structure of a hierarchical society. The internal-combustion engine is the preferred power-pack of an individualizing culture; once produced, moreover, it comes to dominate and to reproduce the consciousness of the epoch, typified in freeway gridlock — "all together alone."

The massive inertia in the infrastructure of automobilism will condition all of our lives indefinitely — suburbs, edge cities, malls, ghettos, gated enclaves, the Sixth Fleet. This is not to say that technologies are a sufficient condition for the shaping or foreclosing of the future. It is salutary to note that the previous largest infrastructural project in U.S. history, the railroad system — which primed the long nineteenth-century wave of capital expansion — lies more or less derelict, all within two lifetimes. There is, finally, nothing self-determining about technology; *contra* the absurd slogan that if something can be made, it will be, "innovation" is continually being blocked by capital flight, state policy, and corporate patent-blitzing.

The value-slope of a technology is no easy thing to estimate, and in any case there is never closure with respect to consequences or reception. Small wonder that Theodor Adorno and Walter Benjamin could so fundamentally disagree about the popular revolutionary potential of cinema. Above all, the value-slope cannot simply be "read off" the artifact itself.

The Amish have some lessons for Luddites in this matter, being canny about the threat posed by new technologies to the character and rhythms of their small, face-to-face communities. The Old Order Amish home has no electricity from the grid, no radio, no television, and, since 1909, no telephone. There are, however, community phones, located in small wooden "shanties" at a distance from dwellings; these phone have unlisted numbers and often no bell to signal incoming calls. The Amish have not anathematized the instrument itself; the accommodation suggests that the key issue is the phone's *relation* to domestic space, community boundaries, and, very probably, gender and age hierarchies.

Of course, we refuse the Amish's theocratic rationale for a critical appropriation of technology, but we also reject principles of resistance based on humanism, since it falsely supposes that the relation of the technological to the human is essentially external. Human nature is not autonomous with respect to technology. Mary Shelley's fantasy of revenge at the hands of science's monstrous creature — in this case, the result of Doctor Frankenstein's Promethean efforts to bypass the female entirely, by galvanizing life from a charnel house — contains the truth that Engels insisted upon: the constitutive role of our artifacts and material productions. We make our selves as we make the world.

If that is so, and since the new technologies of the virtual life are set to compound the old system of domination with fresh colonizations, then E. P. Thompson's attempt to rescue the Luddites from the enormous condescension of posterity turns out to be no sentimental affair. And no easy task, either. A Hearst newspaper editorial of December 10, 1994, rhapsodized: "Cyberspace has a potential to democratize information in ways that were unimaginable scarcely a decade ago," only to continue: "Despite the futile resistance of gape-jawed Luddites, computer technology is here to stay and quill pens are out." A journalist sent to Sproul Plaza to cover the thirty-year reunion of the Free Speech Movement quoted Mario Savio's incendiary call to action against the technocrat's vision of the university as an assembly plant for human capital: "There comes a time when the operation of the machine becomes so odious, makes you so sick at heart, that you can't take part, you can't even passively take part. And you've got to put your bodies upon the gears, and upon the wheels, upon the levers, upon all the apparatus. And you've got to make it stop." (But the writer could not refrain from a typically condescending gloss: he informs us that Savio "intoned in pure Luddite dudgeon." Indeed.)

Not only that — we propose the widest generalizing from the Luddite critique of the machine, to enable a strategic, even a mythic, connection between the lost struggles of the hand-loom weavers against factory discipline and starvation, and contemporary forms of resistance — against the zoning that denatures life in cities, against the mechanization of birth, against racist surveillance and the criminalization of poverty, against the iron cage of bureaucracy, against state borders and identities, against the programming of the wild, so that our world, and our selves, might yet be made over in the textures of William Blake's imagination:

> . . . & hence
> The Web of Life is woven: & the tender sinews of life created.

*References*
The classic account "from below" of the historical Luddites is E. P. Thompson's *The Making of the English Working Class* (London: Victor Gollancz 1963; 2d ed. with new postscript, Harmondsworth: Penguin 1968). David Noble pushed forward the theorizing of Luddism as a principle — drawing on Geoffrey Sea's Mumfordian analysis of machine breaking — in the three-part essay "Present Tense Technology," in *democracy* (Spring/Summer/Fall 1983), developed in *Progress Without People* (Chicago: Charles Kerr 1994). On communications in particular, see Kevin Robins and Frank Webster, *Information Technologies: A Luddite Analysis* (Norwood, New Jersey: Ablex 1986). A critical, sociotechnical history of information technologies is Brian Winston's *Misunderstanding Media* (Cambridge: Cambridge University Press 1986). My remarks on the technics of empire and commerce quote from Daniel R. Headrick, *The Tentacles of Progress* (Oxford: Oxford University Press 1988). My thoughts on

cybernetics have been shaped by the work of Steve Heims, whose *John von Neumann and Norbert Wiener: From Mathematics to the Technologies of Life and Death* (Cambridge: MIT Press 1980) and collective biography of *The Cybernetics Group* (Cambridge: MIT Press 1991) are models of how to reconstruct the history of science with passionate objectivity. Donna Haraway's work challenges humanist-Luddite critiques of the cyber-theorist's assimilation of the human and the artifactual. Paul Edwards has powerfully analyzed the virtual spaces of the Cold War in *The Closed World: Computers and the Politics of Discourse in Cold War America* (Cambridge: MIT Press 1995). The speculations on virtuality were prompted by conversations with Minoo Moallem and Philip Turetzky, and with T. J. Clark and Whitney Davis in the context of perspectivism in the history of art. The discussion of instant replay and video borrows from my essay in *Propaganda Analysis Review* (Summer 1986), which in turn drew on Erik Barnouw's *Tube of Plenty* (New York: Oxford University Press 1977); see also Ben Keen's essay "Play It Again, Sony" in *Science as Culture* 1 (1987). My criticism of positivist linguistics and informationism is a gesture to the work in aesthetics by Philip Turetzky and to the anthropologist Charles Briggs, whose critique of the fetish of "reference" in twentieth-century linguistics may be found in *Learning How to Ask* (Cambridge: Cambridge University Press 1986) and in his forthcoming book with Richard Bauman. The brief remarks on the way statistical uses of the computer efface interiority and depth in the production of virtual "populations" condense my argument made in "The Rhetoric of Risk," *PsychoCulture* 1 (1995). On the question of culture and telecommunications in general, two essential books are Armand Mattelart, *Mapping World Communication* (Minneapolis: University of Minnesota Press 1994), and Jesus Martin-Barbero, *Communication, Culture, and Hegemony* (London: Sage 1993). Appropriation of technology by the Amish is discussed in Richard Sclove's forthcoming book on technology and democracy (New York: Guilford 1995) and in Diane Zimmerman Umble's "The Amish and the Telephone: Resistance and Reconstruction," in Roger Silverstone and Eric Hirsch, eds., *Consuming Technologies: Media and Information in Domestic Spaces* (London: Routledge 1992).

*Herbert I. Schiller*

# THE GLOBAL INFORMATION HIGHWAY:
# Project for an Ungovernable World

W hile plans abound and steps are taken to construct the electronic information highway as rapidly as possible *inside* the continental boundaries, the design for a global system is hardly overlooked. In March 1994, Vice President Al Gore traveled to Buenos Aires, the site of an International Telecommunications Union Conference, at which representatives from 132 countries were present.

Before an international audience, Gore repeated the great promise he saw in electronic communication: "we now have at hand the technological breakthroughs and economic means to bring all the communities in the world together. We now can at last create a planetary information network that transmits messages and images with the speed of light from the largest city to the smallest village on every continent."

After offering this transcendental vision, Gore concluded more pragmatically. This information network, he stated, "will be a means by which families and friends will transcend the barriers of time and distance . . . and it will make possible a global information marketplace, where consumers can buy and sell products."

Buying and selling information goods on the global network, no less than on the domestic one, Gore insisted, required that private industry be in charge. "We propose that private investment and competition be the foundations for development of the GII [Global Information Infrastructure]." (Gore 1994; see also Nathaniel C. Nash, "Gore Sees World Data Privatizing," *The New York Times*, March 22, 1994).

Months earlier, a somewhat different emphasis was given to the electronic communication project in a domestic venue. In September 1993, the White House described the information superhighway as a means "to enable U.S. firms to compete and win in the global economy" and to give the domestic economy a "competitive edge" internationally (NII 1993).

Not mentioned in Buenos Aires, but clearly in the forefront of White House thinking, is the global command and direction of the world economy, through information control, and the benefits that flow therefrom.

This is hardly a recent ambition, expressed for the first time by a new administration. For more than half a century, beginning while World War II was still being waged, U.S. leadership recognized the centrality of information control for gaining world advantage. Well before most of the world could do much about it, U.S. groups, private and governmental, were actively promoting information and cultural primacy on all continents. The policy had many elements in it, not all of them necessarily deliberate.

In the 1990s, for example, when a new generation of leadership is coming into authority around the world, it is common to discover that this or that individual, taking over the reins in this or that country, has been educated in the United States. Mexico provides a striking case in point: its departing president is a Harvard graduate, its newly designated one a Yale product. The education and training of foreign students in American schools and universities, expanded greatly after World War II, are now producing their harvest of graduates who assume high office at home. They have imbibed free market and other doctrines and values ladled out in America's premier schools.

But U.S. global cultural influence has not been limited to formal education. U.S. films and TV programs are the chief fare of national systems in most countries. News programs, especially CNN, offer U.S. perspectives, sometimes the only perspective provided, to world audiences. U.S. recorded music, theme parks, and advertising now comprise a major part of the world's cultural environment.

No less remarkable is the ad hoc adoption of English as the world's second language, facilitated by the waves of U.S. pop culture that have washed across all frontiers for forty years. And once the preeminence of English had been established, Anglo-American ideas, values, and cultural products generally have been received with familiarity and enthusiasm.

All this is well known and amply documented, though the domestic media and political establishments are shy about acknowledging their de facto cultural domination of what they like to refer to as "the global market." What is of special interest here, however, is the skillful combination of information instrumentation with philosophic principle — a mix that fuels the push toward concentrated cultural power. Not the laws of chance but strategic planning, rarely identified as such, underlies this development. It has succeeded well beyond the initial expectations of its formulators.

At the outset of what some hoped would be an American century (Schiller 1992), a vital doctrine was promoted: the free flow of information. Considered out of context, this principle seems unexceptional and indeed, entitled to respect. Yet when viewed alongside the reality of the early postwar years, it conferred unmatchable advantage on U.S. cultural industries. No rival foreign film industry, TV production center, publishing enterprise, or news establishment could possibly have competed on equal terms with the powerful U.S. media-entertainment companies at that time. And so it has gone to this day. The free flow of information, as implemented, has meant the ascendance of U.S. cultural product worldwide.

Under the "free-flow" principle, U.S. global strategy supported the rapid and fullest development of transport and information technologies, which underpinned the capability for the cultural domination that was being constructed. For example, most of the civilian airliners in operation in most countries are U.S.–made. These vehicles have enabled the massive growth of the world tourist industry, which has in turn leaned heavily on U.S. modes of entertaining and nurturing tourists — chain hotels, packaged tours, constructed spectacles, and so on.

Two key sectors received special attention and unstinting resources from the U.S. government in the never-ending pursuit of winning and holding the global market for U.S. products and services: satellite communications radically improved telecommunications and suppressed distance as a factor in global production, and computerization has become the basis of the information-using economy. Both have long been the recipients of heavy subsidies and favored treatment, Washington's enthusiastic rhetoric for "free markets," notwithstanding.

Current plans to construct an information superhighway closely follow the historical model of the U.S. development and deployment of the communication satellite. The satellite project had a single unambiguous goal: capturing control of international communication circuits from British cable interests. The imperial rule of Great Britain in the nineteenth and early twentieth centuries had been facilitated greatly by control of the underwater message flow between the colonies and London. The American-built and -controlled satellite bypassed the cable and helped break the empire's monopoly on trade and investment and to reduce the British role in international communication (Schiller 1992).

Control of information instrumentation invariably goes hand in hand with control of the message flow and its content, surveillance capability, and all forms of information intelligence. To be sure, the revenues from such control are hardly afterthoughts in the minds of the builders and owners of the information superhighway.

Yet again, the "promise" of the communications instrumentation represented by the GII stresses the general social benefit. However, the conditions attached to the proposal — private creation and ownership — make it inevitable that the network will be of greatest value to those who have the financial ability to satisfy their need for instantaneous and voluminous global message flows.

These "information users" are none other than transnational corporations. They constitute the driving force for the creation of a global marketplace, for a deregulated world arena, and for global production sites selected for profitability and convenience — which are also the central considerations behind the National Information Infrastructure (NII) and the GII.

The launching of the global information superhighway project comes at a time when most of the preconditions for a corporate global "order" are in place. There is, first and foremost, the actual existence of a global economy, organized and directed by a relatively tiny number of transnational corporations. According to a survey of this global economic apparatus, 37,000 companies comprise the system. Given that there are millions of businesses in the United States, not counting those in Germany, Japan, and elsewhere, the extent of the concentration of economic influence in this global system cannot be overstated.

And though 37,000 companies occupy the command posts of the world economic order, the largest 100 transnational companies, in 1990, "had about $3.2 trillion in global assets of which $1.2 trillion was outside their own home countries" (*World Investment Report* 1993). These few megafirms are the true power-wielders of our time.

This world corporate order is a major force in reducing greatly the influence of nation states. As private economic decisions increasingly govern the global and national allocation of resources, the amount and character of investment, the value of currencies, and the sites and modes of production, important duties of government are silently appropriated by these giant private economic aggregates (Barnet and Cavenaugh 1994).

These corporations are the leading force in promoting deregulation and privatization of industry in all countries, notably but not exclusively in the telecommunications sector. In a 1994 report *Business Week* notes, for example: "From Italy to Taiwan, scores of governments, caught up in a free market frenzy and needing cash, are selling shares of state-run companies to the public. This sweeping global privatization movement, involving more than 50 nations and expected to raise some $300 billion over five years. . ." (*Business Week*, April 18, 1994).

One effect of the large-scale deregulation of industry and the massive privatizations is the increasing ineffectiveness of national authority. Unaccountability of the transnational corporation is now the prevailing condition in most countries. The world-active company makes fundamental decisions that affect huge numbers of people but reports to no one except its own executives and major shareholders.

At the same time, the strength of the transnational sector continues to grow, and the companies comprising it are themselves engaged in uninterrupted expansion and concentration. These developments are especially notable in the communication-media sphere, which, naturally, is also the site of the strongest sentiments in favor of the information superhighway. As U.S. media and cultural product flow more heavily into the global market, the interests of this sector become increasingly congruent with general transnational corporate objectives and policies.

While nonmedia companies — oil, heavy equipment, aerospace, agribusiness, and others — seek ever-improved means of communication to carry on and extend their international operations, the media-communication sector is only too happy to make these facilities available, at a price. For, of course, this sector it strives to expand markets for its own specific outputs.

The recent rush to integration in the media-communication sector is itself a remarkable development. What can only be described as *total communication* capability — sometimes called "one-stop communication" — has become a short-term goal of the major firms in this sector. This translates into giant companies that possess the hardware and software to fully control messages and images from the conceptual stage to their ultimate delivery to users and audiences.

In brief, what is intended is the creation of private domains that will produce data and entertainment (films, interactive TV programs and video games, recordings, news), package them, and transmit them through satellite, cable, and telephone lines into living rooms and offices. Which companies ultimately will dominate the world and domestic markets is still uncertain. Time Warner, Viacom, Hearst, Bell Atlantic, Sega, US West, Microsoft, AT&T, IBM, Comcast, Tele-Communications, Inc., are a few of the big players that are experimenting with different systems of "full-service" communication and vying with each other for advantageous market position (Ken Auletta, "The Magic Box," *The New Yorker*, April 11, 1994).

"The long-term economic opportunities" of these activities, Ken Auletta points out, "excite the business imagination for the rewards can be stupendous." The sums involved warrant this excitement:

> The cable and telephone businesses today [1994] generate close to two hundred billion dollars a year in revenues. Shopping by catalogue and other forms of shopping at home now constitute an eighty-billion-annual business. Entertainment — video stores, movie theatres, theme parks, music, books, video games, theatre, gambling — is a three-hundred-and-forty-billion-dollar business, and is growing twice as fast as over-all consumer spending.

And this is only the *domestic* market, which continues its expansion! The dimensions of the global market are yet to be discovered. In any event, corporate communication-entertainment titans are readying their capabilities to fill the cultural space of hundreds of millions of users and viewers, at home and abroad. More disturbing still, these global enterprises hope to enjoy and extend their relative immunity to oversight, locally and internationally.

The information superhighway is being promoted as a powerful means to even out the disparities and inequalities that afflict people inside the United States and throughout the world economy. In their many statements about the information superhighway, Vice President Gore and President Clinton insist that the project will reduce the gulf that separates the haves from the have-nots in education, health, and income. But the very basis, the nonnegotiable foundation of the project contradicts that promise.

A privately constructed and owned electronic information system will, of necessity, embody the essential features of a private enterprise economy: inequality of income along with the production of goods and services for profit. As production and sales are inseparably connected to income, the overall economy is directed, by the logic of market forces, to producing for and seeking out customers with the most income — because this strategy offers the greatest possibility of profit. It follows that a privately owned and managed information superhighway will be turned toward the interests and needs and income of the most advantaged sectors of the society. Significant modification of this systemic tendency requires the pressure of a strong political movement.

The most developed countries all exhibit wide income inequalities, and the United States is no exception. In the U.S., the top income levels, representing a tiny fraction of the population, receive more than the amount paid to half of the wage earners.

In less-industrialized countries, most of them still in some sort of economic dependency, the differentials are wider still. In India, for example, 120 million people now enjoy middle-class incomes, but 70 percent of the country's population remain mired in poverty ("As Prosperity Rises, Past Shackles India," *The New York Times,* February 18, 1994).

A recent report (James Brooke, "Colombia Booms Despite Its Violence," *The New York Times,* February 10, 1994) noted that in Colombia

The number of people living below the poverty line has increased by about one million since 1990, to include about half of Colombia's population of 33 million people. In 15 years, the gap between average rural and urban incomes doubled.

This widening gap occurred in a period of "growth."

In a 1994 Human Development report of the United Nation's Development Program, wide income gaps between sections of a nation's population are seen as widespread around the world and threatening chaos in the afflicted areas. Egypt, South Africa, Nigeria, and Brazil, among others, "are countries now in danger of joining the world's list of failed states" (Paul Lewis, "U.N. Lists 4 Lands at Risk Over Income Gaps," *The New York Times,* June 2, 1994).

Similar growth with immiseration is found in the United States. "Today," writes an economic reporter for the *New York Times,* "the economy can keep on growing with the wealthiest 40 percent of the nation's families getting 68 percent of the income, even though 60 percent of the population is unhappily on the sidelines" (Louis Uchitelle, "Is Growth Moral?" *The New York Times Book Review,* March 27, 1994).

Will the creation of privately financed and privately owned, high-speed, multicapability circuits carrying broad streams of messages and images reduce the gaps in living conditions across the globe? Time Warner, AT&T, Microsoft, and their rivals cannot be preoccupied with social inequality. Their focus is on revenue. Profits can come only from those who already have the income to purchase the services that are being prepared for sale.

Failing major political interventions — hardly to be expected in a time of worldwide deregulation and political conservatism — income gaps will widen, not diminish, at home and abroad. The inevitable corollary in communication is the employment of the electronic circuitry for transnational marketing, internal corporate operations, and the ideological objectives of businesses. Corporate data flows, Hollywood films and TV programs, business statistics, home gambling, video games, virtual reality shows, and shopping channels are the likely fare on the new electronic circuitry.

Yet there is at least one cloud on the market-forces horizon: *the question of how this corporately organized world will be governed.* If national authority continues to decline and corporate resource-allocation and general decision making continue to grow, and the welfare of approximately two-thirds of the world's population goes unattended and even deteriorates, what will prevent these conditions from provoking large-scale political convulsions in one place after another? And how can the globally privileged, wherever they may be, insulate themselves from these inevitable upheavals? What authority can check these powerful centrifugal currents?

These matters do not come up regularly on the nation's talk shows, concerned as they are with Madonna's underwear or the suicides of pathetic rock stars. Yet some attention is paid to these issues in more rarefied locales — the cozy diplomatic and foreign policy establishments, private and governmental, that generate the initiatives that eventually become national foreign policy.

In this era of eroding national authority, it is not surprising that some policy formulators have rediscovered the United Nations. In existence for half a century, and bypassed by its most powerful member for most of that time, the UN has, as a result of the changed international scene, drifted back into at least a blurry focus for some national influence-wielders. Yet since its inception, the UN has been a problem for U.S. diplomats. It is an *inclusive* body made up of representatives of all but a very few of the nations in the world, a circumstance that makes it difficult for a few still imperially minded societies to keep 185 national voices from having their say in governing the world.

The original design of the UN aimed to overcome this "obstacle" by putting major decision authority in the Security Council, which is dominated by a small club of the most powerful states. Still, this too proved an unacceptable limitation on U.S. postwar aims, especially given the presence on the council of a rival nonmarket society armed with atomic weapons.

In the early days of the UN there was almost constant deadlock. The creation of NATO (North Atlantic Treaty Organization) and other regional alliances served by design to deprive the one truly international organization of its global role and importance.

The former Soviet Union's frequent blockage of U.S. goals received heavy U.S. media attention, which generally obscured the source of disagreement — U.S. unwillingness to allow the loss of any part of the world market economy. Instead, the fraught messages and scary headlines helped convince Americans of Russian unreasonableness and the need for a gigantic arms program.

The UN was deemed an unworkable organization by U.S. leaders for other reasons as well. The most important objection, rarely made explicit, was to the presence of a large bloc of nations, the former colonial territories. These states, at least in their early postindependence years, constituted a vocal opposition to U.S. and Western efforts to retain or reimpose economic and cultural arrangements that perpetuated these nations' dependency.

The clash of interests between the few highly industrialized and powerful states, with the United States acting as militant whip, and the overwhelming majority of have-nots was epitomized in the struggle, first, for a New International Economic Order (NIEO) and, soon after, a New International Information Order (NIIO). The West refused to allow changes in the prevailing world economic and cultural patterns that favored and perpetuated their interests.

The lesson Washington took away from these engagements in the 1960s and 1970s was to regard the United Nations as an oppositional force. The American media presented the UN as a body inimical to the American way of life. Within a short time, polls would demonstrate convincingly that the American public wanted nothing to do with the UN, and this "public opinion" became the justification for Washington's further anti–UN behavior. It was also a good example, so dear to the hearts of TV executives, of "giving the people what they want" — *after* they have been repeatedly "informed" by the information managers.

The collapse and disappearance of Soviet power made changes inevitable in this long-cultivated popular outlook. A United Nations with a supplicant Russia, instead of a veto-exercising superpower, is more attractive to Washington strategists. Yet the will to dominate without an international mediating body remains strong in some sections of the governing elite.

Elite opinion is split between the strategy of using the United Nations as an instrument for advancing U.S. national interests and the pursuit of an uncompromising unilateralist position. Both perspectives assume that state power will remain in place and continue to prevail, an assumption that gives an air of unreality to both tendencies. Promoters of increased U.S. interest in a Soviet-free UN see that as the best way of organizing and stabilizing the world in accord with the market system. In this view, the post–Cold War UN is a useful extension of U.S. policy. The opposing view believes that state power remains decisive in international affairs — and thus the U.S. should not yield its unilateral control to an international body.

The president and his secretary of state, the top command of the foreign-policy establishment, are deeply attached to the "leadership syndrome." From the beginning of his administration, President Clinton has emphasized, "we are, after all, the world's only superpower. We do have to lead the world" (*The New York Times*, April 24, 1993). And Secretary of State Warren Christopher has chimed in: "I think our need to lead is not constrained by our resources . . . I think that where we need to lead . . . we will find the resources to accomplish that" (Steven A. Holmes, "Christopher Reaffirms Leading U.S. Role in World," *The New York Times*, May 28, 1993).

One glaring weakness of the "leadership syndrome" is Christopher's confident assertion of the resource capability of the U.S. to "lead" the world. This claim is precisely what is called into question by a Rand Corporation analyst. To be a successful hegemonic power, he writes,

> is a wasting proposition. A hegemonic power forced to place such importance on military security must divert capital and creativity from the civilian sector, even as other states, freed from onerous spending for security, add resources to economically productive investments. As America's relative economic strength erodes, so does the comparative advantage over other powers upon which its hegemony is founded. . . . It is difficult to see, therefore, how capitalism can survive the decline of the Pax Americana. (Benjamin C. Schwarts, "Is Capitalism Doomed?" *The New York Times*, May 23, 1994)

A still greater problem confronts those who look forward to a long era of U.S. world "leadership": the reduced capability of all political formations, state, local, regional, to manage, much less control, the vast private economic forces that now are embodied in the transnational corporate system. (We leave aside here the impact on national governance of the renewed strength and clamor of nationalistic and ethnic forces in many parts of the world. These feed on the economic chaos produced by the global market system.)

Early on in his administration, President Clinton outlined succinctly the features of the present world order and some of the dilemmas they produced (speech at the American University, February 26, 1993; text in *The New York Times*, February 27, 1993):

> Capital clearly has become global. Some three trillion dollars race around the world every day. And when a firm wants to build a new factory, it can turn to financial markets now open 24 hours a day from New York to Singapore. Products have clearly become more global. Now, if you buy an American car, it may be an American car built with some parts from Taiwan, designed by Germans, sold with British-made advertisements — or a combination of others in a different mix.

The president elaborated on this transnational corporate scenario:

> Services have become global. The accounting firm that keeps the books for a small business in Wichita may also be helping new entrepreneurs in Warsaw. And the same fast food restaurant that your family goes to — or at least I go to — also may be serving families from Manila to Moscow, and managing its business globally with information, technologies and satellites.

Clinton noted at least a trace of the effects of the operations of this private system:

> Could it be, that the world's most powerful nation has also given up a significant measure of its sovereignty in the quest to lift the fortunes of people throughout the world?

The answer is yes — but not for the reason Clinton offers. "Lifting the fortunes" of people around the world is hardly the motivation of the global corporate system that is reducing the authority of governments everywhere.

Still, from this partial understanding of the workings of the global economy, the Clinton White House has come to a fundamental conclusion concerning the role and importance of information in the routines and practices of the economic order. Here, too, the president's grasp of the new reality commands attention: "Most important of all, information has become global and has become king of the global economy." It follows that "In earlier history, wealth was measured in land, in gold, in oil, in machines. Today, the principal measure of our wealth is information: its quality, its quantity, and the speed with which we acquire it and adapt to it...."

It is this assessment that explains the genesis of the Clinton administration's preoccupation with and support for the new electronic information infrastructure. It is the vast information capabilities that the new infrastructure will provide that excites the government and that prompts the presidential assertion that mastery of this technology will enable the U.S. "to win in the 21st century" (NII 1993).

The reasoning is straightforward. If, in fact, information has become the vital element in the world and domestic economy, the expansion of information capability must confer increased, even uncontested authority on those who have it. This conclusion reinforces the unilateralist position. Why offer support to the United Nations or any other international body if the means of global authority — information control — are at hand? But is it so simple?

Those who believe state power will be enhanced by the new information technologies and expanded information flows may be overlooking one critical point. The main beneficiaries of the new instrumentation and its product are likely to be the transnational corporations. They will always be the first to install and use these advanced communication technologies.

The strength, flexibility, and range of global business will become more remarkable. The capability of the state, including the still very powerful United States, to enforce its will on the economy, domestic or international, will be further diminished. This may be partly obscured for a time because the national security state will have at its disposal an enhanced military and intelligence capability, derived from the new information technologies.

Interest rates, capital investment, employment, business-cycle policy, local working conditions, education, and entertainment increasingly elude national jurisdiction. Creation of a far-flung information superhighway will accelerate the process. This suggests that the government's information policy is a recipe for further diminution of national power, one that will encourage even greater concentration of private, unaccountable economic influence in geographically dispersed locales.

Some people see this corporate undermining of state power as a development to be encouraged. Given the long history of coercive state power, this view is certainly appealing. But politics must take account of prevailing power relationships: while the national state remains a potentially repressive force, today private, unaccountable economic power constitutes a greater threat to individual and community well-being.

The contours of the world-in-the-making, of progressively enfeebled governments, are shadowy, but one description, heavily influenced by the role of the new information technologies, is offered by Alvin Toffler. Toffler, one of the early boosters of the information-using economy and, accordingly, a darling of the speculative financial community, has long called the current historical epoch "the Third Wave" (Toffler 1980). In his typology, this denotes the shift first from agricultural to industrial society and now to an information-using society.

According to Toffler, global organization, production, distribution, work, living arrangements, and war itself have been profoundly affected. More cognizant than most of emerging realities, Toffler foresees the development of global "niche economies." Though he doesn't define them as such, niche economies can only be understood as enclaves of successful, transnational corporate activity. Some of the sites Toffler mentions include regions in southern China, parts of the former Soviet Union, the Baltic states, and southern Brazil — highlighting the fact that the sharp divisions between the well-off and the disadvantaged occur *within* as well as between countries.

Though inequality in the social order is nothing new, capitalist "development" accentuates and deepens this condition. And the new information technologies extend inequality by providing additional capabilities — mobility, flexibility, instantaneity — to the global corporation.

In *The Global City,* Saskia Sassen views the rise of what she calls global cities — Tokyo, London, New York, and lesser centers — as the direct outcome of the operations of the transnational corporations. These giant firms require a wide range of what Sassen calls "producer services" — advertising, design, accounting, financial, legal, management, security, and personnel — which can be concentrated in a few metropolitan centers. The life of these new centers may offer a glimpse of the future for some parts of the world population (Sassen 1991).

For Kenichi Ohmae, writing in the *Wall Street Journal*, the future is already here:

> No longer will managers organize international activities of their companies on the basis of national borders. Now the choice will be not whether to go into, say, China, but which region of China to enter. . . . The primary linkages of these natural economic zones are not to their 'host' countries but to the global economy.

Ohmae finds that the best example of what he is describing

> is Dalian, a prosperous city of 5.2 million people in Liaoning Province in northern China. Dalian' s prosperity has been driven not by clever management from Beijing but by an infusion of foreign capital and the presence of foreign corporations. Of the 3,500 corporations operating there, as many as 2,500 are affiliates of foreign companies from all over the world. . . . In Dalian you can virtually smell the global economy at work (Kenichi Ohmae, "New World Order: The Rise of the Region-State," *The Wall Street Journal*, August 16, 1994).

Toffler's "niche economies" also contribute to what he terms "the Revolt of the Rich." In the past, it was invariably the poor who revolted against the rich. But now well-off groups and locales want to preserve and extend their advantages. They do their best to distance themselves from, and to discard, their lagging and disadvantaged countrymen, regions, states.

Toffler's explanations may be deficient, but his descriptions are accurate enough. The poor, a good part of the minority population, and the inadequately educated are increasingly cordoned off in urban centers, jails, hospitals, and isolating areas and institutions.

Similarly, privileged countries today try to seal themselves off from masses of desperate people who wish to escape from destitute home areas. Western Europe tries to keep out the Africans and the East Europeans, Japan maintains tight control over immigration, Washington is wary of the human tide from the Southern Hemisphere that presses against the continental borders.

Can the rich enclaves, favored groups, and still relatively viable nation-states succeed in severing their ties with their poor neighbors, inside and outside their borders? While Toffler doesn't directly address this central question, he does speak about "niche war," making the connection clear. He regards the Persian Gulf War as an early model of what may be in store for those seeking to challenge the new world corporate order. Toffler puts it this way:

> [I]f we are now in the process of transforming the way we create wealth, from the industrial to the informational . . . there is a parallel change taking place with warfare, of which the Gulf War gives only the palest, palest little hint. The transition actually started back in the late 1970s, early 1980s, to a new form of warfare based on information superiority. It mirrors the way the economy has become information-dependent.

Toffler further predicts that

> In military terms there will be attempts to coordinate all the knowledge-intensive activities of the military from education and training to high-precision weaponry to espionage to everything that involves the mind — propaganda — into coherent strategies. ("Shock Wave Anti-Warrior," *Wired,* November 1993)

Along with the "niche war" strategy to overcome social eruptions, Toffler endorses the transmission of news and information to disaffected areas. CNN, for example, is regarded favorably as a suitable channel in such endeavors, as are the BBC and Japan's NHK. These networks are seen as unproblematic, reliable vehicles, dispensing "news" and information that will undermine the dissidents, whoever they may be.

Washington's plan for an information superhighway does not mention these applications, but the thinking underlying the project surely takes them into account. What else can it mean when the installation of the new information technologies are regarded as the vehicle to "win in the 21st century"?

But can precision warfare with a high information component and control of global news flows keep the world orderly while privately initiated economic forces are contributing to wildly disproportionate income distribution and gravely distorted resource utilization, locally and globally?

The deepening crisis that is provoked by advanced technology, used mainly for corporate advantage and implemented according to the rules of the market, may summon forth even less promising "solutions." Direct military interventions in nations "where governments have crumbled and the most basic conditions for civilized life have disappeared . . . is a trend that should be encouraged," writes one historian whose views have had respectful attention in the mainstream media.

According to this writer, the root cause of problems in many Third World countries, especially African nations,"is obvious but is never publicly admitted: some states are not yet fit to govern themselves" (Paul Johnson, "Colonialism's Back — and Not a Moment Too Soon," *The New York Times Sunday Magazine,* April 18, 1992). But it is not a matter of being fit to govern oneself, a patronizing, if not racist charge. In the last years of the twentieth century, satisfactory governance is in crisis almost everywhere. The crisis derives from the weakening of state authority that has been brought about by half a century of Cold War conflict, in tandem with the expansion of unaccountable private economic power. Information technologies at the disposal of this power further exacerbate already bad conditions. The response to this global crisis demands a totally different economic, political, and cultural direction from what now prevails.

*References*

Barnet, Richard, and Cavenaugh, John. 1994. *Global Dreams.* New York: Simon & Schuster.

Gore, Albert. 1994. Remarks delivered at the meeting of the International Telcommunications Union, Buenos Aires (March 21).

The National Information Infrastructure (NII). 1993. Agenda for Action, Executive Summary. Washington (September 15).

Sassen, Saskia. 1991. *The Global City.* Princeton: Princeton University.

Schiller, Herbert I. 1992. *Mass Communications and American Empire.* New ed. Boulder: Westview.

Toffler, Alvin. 1980. *The Third Wave.* New York: William Morrow.

World Investment Report. 1993. Transnational Corporation and Integrated International Production. United Nations Conference on Trade and Development, Programme on Transnational Corporations. New York: United Nations.

*Oscar H. Gandy Jr.*

# IT'S DISCRIMINATION, STUPID!

A lthough each day seems to bring us news about yet another electronic marvel that promises to "free us" from drudgery, long lines, and insolent service workers, we are just beginning to hear about the goodies that are in store for us once we make our way onto "the information superhighway." The responsibilities of the vice presidency and the availability of a larger pool of speechwriters may have served to transform Al Gore's numbing rhetoric about the National Information Infrastructure into more colorful and popularized marketing buzzwords, but they haven't changed the fact that all this handwaving about the wonders of the Information Age is an attempt to sell us a bill of goods.

To be sure, there is much benefit to be derived from the use of information technology. The corporations that use these technologies first gain short-term competitive advantage against others that lag behind. Individual users may also gain time and avoid annoyances in many of the routine transactions that make up their day-to-day lives. Examples of these gains fill the promotional brochures and advertisements at the same time that they slip into the cultural mainstream of news and entertainment. The marvels of AT&T's telecommunications future seem downright liberating, although nagging doubts about the implications of always being "in touch" may trouble some viewers. What is largely missing is the bad news, awareness of the downside consequences that doesn't diffuse as readily because nobody pays a public relations flack to spread the word about the unemployment and widespread economic discrimination that the use of this information technology almost guarantees. That task is left to the technorealists, or contemporary Luddites like myself, who continually seek to introduce a word of caution about the latest technological "advance."

Most recently, I have been writing and speaking about what most people associate with concerns about privacy. But really, it is not the loss of privacy that concerns me. Rather I hope to raise the general level of awareness of the economic and political consequences that flow from the loss of control over personal information. Government actions to protect privacy are quite imperfect responses to the problems of discrimination that the collection, sharing, and use of personal information help create. The choice of privacy as the entry point is a strategic rather than a fundamental value position. It reflects an assessment of the difficulty of pursuing what are essentially egalitarian goals through a discourse on discrimination that has been distorted by recent debates about civil rights and a right-wing counterattack against the specter of political correctness. This conflation of issues was best reflected in the effective mobilization of popular opinion against Bill Clinton's nominee for civil rights, Lani Guinier.

By talking about privacy, however, I seem to strike a sensitive nerve. Public opinion surveys reveal that most people are very concerned about privacy. Yet, curiously, these same surveys also indicate that most people are unable to identify incidents when their own privacy had been invaded. However, even though they may be unaware of it, I have no doubt that the quality of their lives has been changed because some organization has made a decision on the basis of information about them or about someone "like" them. These decisions take place in the background, outside our conscious awareness, and they make use of a technology that is generally not well understood. I refer to this technology as the "panoptic sort." It is a discriminatory technology that assigns people to groups of winners and losers on the basis of countless bits of personal information that have been collected, stored, processed, and shared through an intelligent network.

Because this process takes place in the background, this very real sense of an evaporating zone of privacy has been produced and reinforced primarily by a number of news stories that talk about surveillance in different spheres of our lives. We read quite a bit about surveillance in the workplace. A recent study by the International Labor Organization (ILO) identified American workers as being subject to the most intense surveillance pressure of any population of laborers they surveyed. Networked computers facilitate the "remote sensing" of workers by capturing and comparing keystrokes, e-mail communications, completed transactions, and even the tone of voice of telemarketers, agents, and inbound claims processors. In the face of rising medical costs, employers have actively sought to extend the scope of surveillance to include activities and interests pursued off the job that may have implications for employee health. Because of increased vulnerability to suits for negligent hiring, as well as rising costs of worker training, surveillance of American workers is also being front-loaded, where background checks are part of comprehensive preemployment screening.

News stories have also served to maintain a relatively high level of concern about surveillance by the government. Government surveillance of dissidents suspected of political or ideological crimes is being replaced by surveillance of individuals and corporations suspected of economic crimes related to unauthorized foreign trade, money laundering, or trade in illegal substances. The use of commercial databases as aids in the identification of tax avoiders or in the general pursuit of fraud, waste, and abuse contributes to a heightened sense of distrust. Distrust of the government's promises regarding the confidentiality of the data resulted in a record level of refusals to participate in the 1990 census. Press coverage of the often arcane debates about the government's plan to ensure access to "secure" encrypted communications through the use of a "Clipper chip" with an "escrowed" key that the government could obtain, just adds fuel to the fire.

To a more limited extent, a number of mass-market books by journalists have raised the level of public awareness about the ways in which privacy is invaded by corporations seeking to improve the efficiency of their marketing campaigns.

Yet none of this writing frames the problem of privacy and surveillance in terms of discrimination, which is the end result of the social construction of difference in the pursuit of profit and social control. The panoptic sort, as a "difference engine" in support of rationalization and efficiency, is not limited to any single sphere of our existence. Personal information is used to determine our life chances in our roles as citizens as well as in our lives as employees and consumers. Techniques developed in one sphere migrate rapidly into use in other spheres. With the increase in privatization and with the elevation to the status of gospel of the "partnership" bargain between business and government, the boundaries between spheres is in fact dissolving. However, because the discourse on privacy and surveillance has largely ignored discrimination within markets, I will focus my attention there.

### Marketing and the Panoptic Sort

Often when I speak about the threats to freedom and privacy that I associate with the use of personal information for marketing, I "go for the gasp" by telling people about the kinds of information that are readily available in the largely unregulated marketplace. Reading the magazines and newsletters that serve the direct-marketing industry throws up new examples of commercial lists that can be counted on to raise some eyebrows and occasionally the blood pressure of some of the folks in the audience.

Many are unaware that there is a multimillion-dollar industry that serves direct marketing and other kinds of opportunity-seeking risk-avoiding organizations by renting access to computer files or lists of individuals derived from nearly every moment of contact with an organization. Not only purchases but applications and general inquiries generate records that have potential value to an organization interested in classifying an individual as a member of a particular group or market segment. Membership in such a segment implies a greater susceptibility to a targeted promotional appeal. In a recent issue of *DM News,* a weekly serving the direct marketing industry, I noted the availability of something called the "National Psychic Network file," which contained some 30,000 individuals who had made calls to the network in the last three months "wishing to speak to their own psychic advisor." They called for answers to questions about love, money, and other decisions they were currently facing. These are individuals who spent $2.99 per minute, with average calls costing between $25 and $50.

If these folks didn't seem sufficiently attractive to scam artists, or others with nonessentials for sale, then there was the "Lift the Ban

Donor" file offered by Strubco, a vendor that specializes in "gay" lists. This particular list includes 2,000 donors who responded to full-page ads in the *New York Times*, 10,000 who responded to direct-mail solicitations, and 21,000 who responded to telemarketing campaigns seeking support for the effort to "lift the ban preventing lesbians and gays from serving openly in the US military." According to Strubco, 75 percent of those who responded to newspaper ads were heterosexual, making contributions averaging $26, while the majority of those responding to the telephone appeals were homosexual and gave more than $40. Although Strubco indicates that these particular lists are not selectable by sexual orientation, many of the other lists it offers are explicit in this regard.

Perhaps a really creative marketer in search of the truly impressionable might decide to do a "match" of one of Strubco's lists against the 200,000-name "American Patriot Donors List." According to Response Unlimited, the group managing the list, "these contributors support conservative causes such as opposing gays in the military, gay rights issues, pornography, government waste, tax increases . . . [and in support of] Oliver North, the Gulf War, and sending bibles to Russia." Lists like these are regularly used for the cold calls or prospecting letters sent by direct marketers seeking to establish a relationship with a new customer. Increasingly, they are being used to enhance the value of information about current customers already in an organization's database. This information is a valuable strategic resource that has been recognized as a depreciable asset for purposes of taxation. Its primary value, of course, is for prediction.

## Predictive Models

The collection and processing of personal information is directed toward the reduction of uncertainty about the average response of a group. The goal is strategic and rational. As a reporter for the *American Banker* discusses it, the marriage of the computer with the telecommunications network makes it cost effective to acquire and process great masses of data so as to extract "actionable consumer intelligence." This intelligence is used to inform decisions about the benefits to be gained from making offers to one group while ignoring other groups entirely.

"Response modeling" is just one of many names for the generalized approach toward maximizing positive returns while minimizing communication-related marketing expenses. The intelligence is represented in a single score that may be used to place an individual along a scale that may be divided into categories indicating a break-even response rate. The rational marketer will communicate with individuals above that point. The data that are used to develop and evaluate the model come from a variety of sources, all of which raise questions about privacy and informed consent. The most troublesome is that which is purchased from third parties, but even the use of information gathered from the behavior of an organization's own customers or clients raises important ethical questions.

In one banking case study, the author noted the use of what he termed "hard exclusions." These exclusions were called "hard" because they represented criteria that "will never be changed." The use of such criteria results in the identification of accounts that "would not be promoted under any condition"; and in this particular case, that represented about half the portfolio of over two million accounts.

The ability to make use of personal information for marketing purposes is naturally correlated with market power. My own research has determined that it is the larger firms that are more likely to develop their own databases and enhance that data with information acquired from third-party vendors. These organizations are also likely to classify themselves as sophisticated users of computers, with data-processing specialists on staff. For organizations without the level of in-house expertise needed to develop and test response or other predictive models, the market provides an ever-expanding number of information specialists who offer assistance in developing models on demand.

Yet it is also clear that this technology is rapidly moving downscale. The furor over the CD-ROM list-generator, "Marketplace: Households," was aimed at a rather low level of technology with the capacity for generating lists from predetermined attributes that might be combined to facilitate intelligent narrowing. But there are a number of software packages being introduced that are aimed at the thousands of nontechnical users of personal computers in midsize firms. One product called DataLogic is supposed to help users to "build, evaluate and implement 'rough sets' models." Capable of using less-than-perfect data, this approach predicts membership in an identifiable group or set. Now, this program can tie up a PC for nearly two hours to process 10,000 records, but the advantage of such an approach to modeling is that the software will determine which variables to eliminate as being irrelevant, redundant, or otherwise useless for assigning individuals to groups.

### So What's the Harm?

Informed by such models, organizations are better able to offer different consumers more- or less-generous terms. Economists suggest that in markets that are less than fully competitive, providers of goods and services can approach the ideal of perfect discrimination where the organization captures not one penny less than what a customer might be willing to pay. The consequence is a market that becomes increasingly inefficient as monopolization expands.

This concern is based on the recognition that multiproduct firms have competitive and strategic advantages over present and future competitors because of their privileged access to customer information. Telecommunications firms like AT&T that are also in the business of granting credit, and which perhaps will soon be in the business of providing information and entertainment through subsidiaries or partnerships, will have a distinct advantage over smaller entities in fewer lines of business. That is, credit information, and the information derived from numerous transactions, is available to the multiproduct firm at a far lower cost than it might be acquired (if at all) by competitors. Most of my concerns, however, are not focused on the market or society, but on the consequences for individuals.

Because the panoptic sort is discriminatory by definition, it represents a problem for the radical egalitarians who deny the gods of efficiency and profit to the extent that institutionalized choices result in differences in the quality of life that each individual is able to enjoy. Because at its best the panoptic sort is guided by a utilitarian, rather than an ethical standard, I have little hope for a policy response that will satisfy my requirements for freely granted, fully informed consent. Self-identified privacy advocates, many of whom believe that the solution to the problems of privacy in the context of commercial use of personal information are to be found in establishing a market for personal information, fail to understand that such markets will be biased and inefficient. These markets must fail for several reasons, but the most important of them include the substantial inequities between buyers and sellers, both in terms of power and in terms of the information that people need to determine and to assign prices to the information rights they are pressured to grant. Unequal markets reproduce inequality.

Traditional concerns about privacy are associated with threats to the autonomous individual. While such a focus is important, and is perhaps necessary if privacy protection is to enjoy the kind of political viability that advocates of national privacy commissions desire, I would like to at least begin thinking about other kinds of vulnerability that are more difficult to engage, both conceptually and in terms of feasible political action. The notion of group privacy is not entirely new. The subject has been raised at the level of geographic communities that have been victimized by erroneous classifications, as in the case of a credit agency labeling an entire community as deadbeats, or when vendors of sexually explicit material sought to use video rental data to demonstrate the level of community standards. The privacy standards developed by Warner-AMEX following this latter case represented an effort to forestall concerns that we would later see being addressed in part by the Video Privacy Protection Act. The concern was that an entire community would be classified in terms of the tastes of some subset of consumers. Yet geographically defined communities are already identified in terms of the average income, education, and spending habits of neighbors within a zip code or census-defined boundary. Classifications of communities on the basis of geodemographic clustering models are used to guide decision making in a wide variety of economic and political areas.

Political sensitivity, not to say political correctness, may limit the use of particular kinds of information as the basis for discriminatory classification. But our concern with discrimination ought not to be limited to groups whose collective economic and political strength allows them to claim some degree of protection by requiring that decisions be made without reference to their unique status as members of a protected group. Through the operation of the panoptic sort, new "groups" are identified every day. Unlike the groups that have traditionally been the focus of study by sociologists and anthropologists, the members of these analytical groups may not actually be aware of their membership or of the rules that determine their inclusion.

The notion of categorical vulnerability, where membership in the constructed class is neither voluntary nor cognized by the persons so objectified, represents a special problem of critical social theory. The literature on class, race, and gender consciousness is filled with talk about the conditions through which such a consciousness is produced. Group consciousness is, of course, important for the recognition of common circumstance and is seen to be a fundamental component of solidarity. Debate within African-American studies centers on the role of shared experience in the development of an Afrocentric perspective. Sociological investigations into the failure of working-class consciousness or solidarity to develop in the context of a service economy point to the importance of transactions in the market for labor as a source of group identification. But when such classification takes place out of view, literally behind the backs of individuals, the basis for the identification of common circumstance is elusive at best.

Even though Howard Rheingold's recent book, *The Virtual Community*, suggests that computer-mediated communication facilitates the identification of common interests, the sharing of information, and even provides a highly efficient means of mobilization, the required level of awareness rarely achieves critical mass. People are classified in terms of their abilities and disabilities, and increasingly in terms of their medical status (for example, people with HIV/AIDS). But when the groups to which people are assigned are the products of multivariate clustering techniques, where the relevant variables number in the hundreds, and the contribution of each variable is indicated by a single coefficient reflecting all the other variables held constant statistically, it is impossible for the individual to understand how to act to modify his or her status.

Increasingly the operation of the panoptic sort is being automated. Rules-based, or expert, systems are used to arrive at determinations about applicants for a variety of services like insurance or health management, where the provider has an incentive to reduce risk by denying applicants or challenging claims. The same kind of programs that process job applications automatically through the use of optical scanners for résumés are used to score and evaluate applications for credit or apartment leases. Resource-management programs that rate and rank incoming telephone calls on the basis of geodemographic information and expected value increasingly determine whether prompt service or a busy signal greets each caller. Caller-ID or other authentication technology supports the automation of such classification.

There is no question that from the perspective of the service provider, the use of automated classification that supports the avoidance of risk is rational and profitable. From the perspective of individuals looking for a job, for housing, for insurance, or even for information to guide a purchase decision, the fact that they were denied service, or a discount, or were assigned to a queue, provides little information about the nature of or the basis for their classification. Indeed, the difficulty of knowing how these different pieces of information contribute to one's classification represents a fundamental challenge to the pursuit of informed consent as a condition of fair information use.

## Informed Consent

Not all the relevant information used in the panoptic sort is provided by an individual in an application or in response to inquiries. In many cases, this information has been gathered indirectly from third parties, or information consolidators, whose business it is to facilitate the construction of consumer profiles. This information is generated from an infinite variety of transactions, many of them automated, as in self-service credit card purchases of fuel, food, and entertainment, but also including actions as innocuous as calls to 800 numbers for information. Because these transactions may involve multiple parties in different organizations, there is considerable debate about who "owns" or has property rights in that information — while the rights of the consumer are rarely at issue. As many as 93 percent of consumers seem to agree that there ought to be legal requirements reflecting the moral responsibility of marketers to seek permission from individuals before they share personal information derived from transactions. Unfortunately, this view is at considerable variance with mainstream business practice, and the courts tend to support the corporate stance. When challenged by privacy advocates, the direct-marketing industry has been fabulously successful in establishing as normal that businesses may use personal information without limit as long as they provide the consumer with the opportunity to "opt out" of the activities that they find troublesome. While there are any number of objections to "opt out" in terms of the burden that it places on the individual, my concern is more fundamental. The legitimation of "opt out" formalizes the automatic grant of broad consent to a whole host of uses of personal information that can be argued to meet a "legitimate business purpose."

"Opt out" is also problematic because, as simple as it is, it is a concept that is apparently easily misunderstood. In testimony before a House committee investigating misuse of the National Change of Address service, a private citizen, who just happened to be a patent attorney, suggested a solution to the problem of unwanted use: "the right way to fix this problem would be to let postal patrons 'opt out.' There would be a box to check, optionally giving the Postal Service permission to provide the new address to everybody. If the box were not checked, the Postal Service would be obligated to use the information on the form only for forwarding." Unfortunately, what this attorney described to Congress, and seemed clearly to prefer, was what the industry understands and dreads as "opt-in," or affirmative consent, where use of personal information would be limited to the purposes for which it was initially granted, unless a more generous grant was explicitly provided. For informed consent to be fully operational, it must be extended to all the ways in which information about individuals is ultimately used in the panoptic sort.

## Market Research

Because medical research involves readily observable risks to individuals, we understand how we eventually developed a routine aimed at ensuring that consent is fully informed and is obtained by noncoercive means. There is also a well-developed ethical position among social scientists that requires informed consent from subjects, and many protocols require the debriefing of subjects after experimental treatment. Of course, the real risks in many of these studies do not occur within the experiment or the data-gathering phase, but occur as a consequence of the social policies that are eventually implemented on the basis of the data that have been gathered. Social scientists haven't gone very far in addressing this aspect of the risks to which research subjects are unwittingly placed, but many do recognize it as an ethical concern.

On the other hand, market research, beyond that undertaken by academics with publication and tenure in mind, almost never involves collecting an explicit grant of consent. With most surveys and interviews, consent is assumed to have been granted as long as the respondent continues to provide answers to questions. The marketing research that is involved in the development and evaluation of response marketing models rarely if ever involves informing the consumer that he or she is participating in a research activity that may have consequences for the quality of services that they may receive.

More problematic from the stance of critical social theory is the fact that the intelligence produced by marketing research is based on representative samples, but it influences the options that are available to millions of others who are not directly measured. There is no way, beyond regulation or authentic public opinion, for the millions who participate indirectly in marketing research to grant or deny their consent, even though they have surely been placed at risk.

## The Future

Perhaps in the electronic futures that are envisioned by futurists and science-fiction writers, the marketplace will move toward the ideal that economists imagine, where buyers and sellers are fully informed, able to negotiate as equals in a network environment where transactions are instantaneous, costless, and anonymous. Perhaps computer scientist David Chaum's "digital money" will help speed us in that direction by providing for autonomous but authenticated purchases. Indeed, the promise of "intelligent agents" searching the marketplace for the highest quality at the best price is really something of a privacy fundamentalist's dream. These "knowbots" would overcome the problems that we associate with human agents who continually have to struggle with conflicts of interest. This would be a market without the need for advertising or other forms of strategic communication that obfuscate and misdirect rather than inform.

Of course, such a future is not assured. At the same time that computer professionals work frantically to develop digital agents and other aids to the exploration of the information environment, we find that the press of commercialization has invaded the Internet and will spread like the worst kind of virus. While we may be awed by the potential of hypermedia to facilitate the searching of the global information super-highway, it seems clear that these navigational services will be tempted to meet their costs, and add to the bottom line by linking advertising copy with particular "cards," in the same way that Prodigy clutters up each of its screens. Cards that don't produce will be dropped, or will be more difficult to find; or more likely, people who don't use the commercially viable paths will be dropped or ignored. Resolute avoiders like me will be faced with reports from our agents that the information we seek is "not available at this time."

In the meantime, while disrupters and anarchists seek to introduce error and noise into the system, and academics seek to raise consciousness through their contributions to the literature of alarm, privacy activists will continue to prepare testimony and electronic press releases. All this activity is necessary because, as Jeff Smith suggests in his recent book, *Managing Privacy,* the privacy-intensive industries are unlikely to pursue meaningful self-regulation on their own.

*Laura Miller*

# WOMEN AND CHILDREN FIRST:
## Gender and the Settling of the Electronic Frontier

W hen *Newsweek* (May 16, 1994) ran an article entitled "Men, Women and Computers," all hell broke out on the Net, particularly on the on-line service I've participated in for six years, The Well (Whole Earth 'Lectronic Link). "Cyberspace, it turns out," declared *Newsweek's* Nancy Kantrowitz, "isn't much of an Eden after all. It's marred by just as many sexist ruts and gender conflicts as the Real World. . . . Women often feel about as welcome as a system crash." "It was horrible. Awful, poorly researched, unsubstantiated drivel," one member wrote, a sentiment echoed throughout some 480 postings.

However egregious the errors in the article (some sources maintain that they were incorrectly quoted), it's only one of several mainstream media depictions of the Net as an environment hostile to women. Even women who had been complaining about on-line gender relations found themselves increasingly annoyed by what one Well member termed the "cyberbabe harassment" angle that seems to typify media coverage of the issue. Reified in the pages of *Newsweek* and other journals, what had once been the topic of discussions by insiders — on-line commentary is informal, conversational, and often spontaneous

— became a journalistic "fact" about the Net known by complete strangers and novices. In a matter of months, the airy stuff of bitch sessions became widespread, hardened stereotypes.

At the same time, the Internet has come under increasing scrutiny as it mutates from an obscure, freewheeling web of computer networks used by a small elite of academics, scientists, and hobbyists to . . . well, nobody seems to know exactly what. But the business press prints vague, fevered prophecies of fabulous wealth, and a bonanza mentality has blossomed. With it comes big business and the government, intent on regulating this amorphous medium into a manageable and profitable industry. The Net's history of informal self-regulation and its wide libertarian streak guarantee that battles like the one over the Clipper chip (a mandatory decoding device that would make all encrypted data readable by federal agents) will be only the first among many.

Yet the threat of regulation is built into the very mythos used to conceptualize the Net by its defenders — and gender plays a crucial role in that threat. However revolutionary the technologized interactions of on-line communities may seem, we understand them by deploying a set of very familiar metaphors from the rich figurative soup of American culture. Would different metaphors have allowed the Net a different, better historical trajectory? Perhaps not, but the way we choose to describe the Net now encourages us to see regulation as its inevitable fate. And, by examining how gender roles provide a foundation for the intensification of such social controls, we can illuminate the way those roles proscribe the freedoms of men as well as women.

For months I mistakenly referred to the EFF (an organization founded by John Perry Barlow and Lotus 1-2-3 designer Mitch Kapor to foster access to, and further the discursive freedom of, on-line communications) as "The Electronic Freedom Foundation," instead of by its actual name, "The Electronic Frontier Foundation." Once corrected, I was struck by how intimately related the ideas "frontier" and "freedom" are in the Western mythos. The *frontier*, as a realm of limitless possibilities and few social controls, hovers, grail-like, in the American psyche, the dream our national identity is based on, but a dream that's always, somehow, just vanishing away.

Once made, the choice to see the Net as a frontier feels unavoidable, but it's actually quite problematic. The word "frontier" has traditionally described a place, if not land then the limitless "final frontier" of space. The Net on the other hand, occupies precisely no physical space (although the computers and phone lines that make it possible do). It is a completely bodiless, symbolic thing with no discernable boundaries or location. The land of the American frontier did not become a "frontier" until Europeans determined to conquer it, but the continent existed before the intention to settle it. Unlike land, the Net was created by its pioneers.

Most peculiar, then, is the choice of the word "frontier" to describe an artifact so humanly constructed that it only exists as ideas or information. For central to the idea of the frontier is that it contains no (or very few) other people — fewer than two per square mile according to the nineteenth-century historian Frederick Turner. The freedom the frontier promises is a liberation from the demands of society, while the Net (I'm thinking now of Usenet) has nothing but society to offer. Without other people, news groups, mailing lists, and files simply wouldn't exist and e-mail would be purposeless. Unlike real space, cyberspace must be shared.

Nevertheless, the choice of a spatial metaphor (credited to the science-fiction novelist William Gibson, who coined the term "cyberspace"), however awkward, isn't surprising. Psychologist Julian Jaynes has pointed out that geographical analogies have long predominated humanity's efforts to conceptualize — map out — consciousness. Unfortunately, these analogies bring with them a heavy load of baggage comparable to Pandora's box: open it and a complex series of problems have come to stay.

The frontier exists beyond the edge of settled or owned land. As the land that doesn't belong to anybody (or to people who "don't count," like Native Americans), it is on the verge of being acquired; currently unowned, but still ownable. Just as the ideal of chastity makes virginity sexually provocative, so does the unclaimed territory invite settlers, irresistibly so. Americans regard the lost geographical frontier with a melancholy, voluptuous fatalism — we had no choice but to advance upon it and it had no alternative but to submit. When an EFF member compares the Clipper chip to barbed wire encroaching on the prairie, doesn't he realize the surrender implied in his metaphor?

The psychosexual undercurrents (if anyone still thinks of them as "under") in the idea of civilization's phallic intrusion into nature's passive, feminine space have been observed, exhaustively, elsewhere. The classic Western narrative is actually far more concerned with social relationships than conflicts between man and nature. In these stories, the frontier is a lawless society of men, a milieu in which physical strength, courage, and personal charisma supplant institutional authority and violent conflict is the accepted means of settling disputes. The Western narrative connects pleasurably with the American romance of individualistic masculinity; small wonder that the predominantly male founders of the Net's culture found it so appealing.

When civilization arrives on the frontier, it comes dressed in skirts and short pants. In the archetypal 1939 movie *Dodge City*, Wade Hatton (Errol Flynn) refuses to accept the position of marshal because he prefers the footloose life of a trail driver. Abbie Irving (Olivia de Haviland), a recent arrival from the civilized East, scolds him for his unwillingness to accept and advance the cause of law; she can't function (in her job as crusading journalist) in a town governed by brute force. It takes the accidental killing of a child in a street brawl for Hatton to realize that he must pin on the badge and clean up Dodge City.

In the Western mythos, civilization is necessary because women and children are victimized in conditions of freedom. Introduce women and children into a frontier town and the law must follow because women and children must be protected. Women, in fact, are usually the most vocal proponents of the conversion from frontier justice to civil society.

The imperiled women and children of the Western narrative make their appearance today in newspaper and magazine articles that focus on the intimidation and sexual harassment of women on line and reports of pedophiles trolling for victims in computerized chat rooms. If on-line women successfully contest these attempts to depict them as the beleaguered prey of brutish men, expect the pedophile to assume a larger profile in arguments that the Net is out of control.

In the meantime, the media prefer to cast women as the victims, probably because many women actively participate in the call for greater regulation of on-line interactions, just as Abbie Irving urges Wade Hatton to bring the rule of law to Dodge City. These requests have a long cultural tradition, based on the idea that women, like children, constitute a peculiarly vulnerable class of people who require special protection from the elements of society men are expected to confront alone. In an insufficiently civilized society like the frontier, women, by virtue of this childlike vulnerability, are thought to live under the constant threat of kidnap, abuse, murder, and especially rape.

Women, who have every right to expect that crimes against their person will be rigorously prosecuted, should nevertheless regard the notion of special protections (chivalry, by another name) with suspicion. Based as it is on the idea that women are inherently weak and incapable of self-defense and that men are innately predatory, it actually reinforces the power imbalance between the sexes, with its roots in the concept of women as property, constantly under siege and requiring the vigilant protection of their male owners. If the romance of the frontier arises from the promise of vast stretches of unowned land, an escape from the restrictions of a society based on private property, the introduction of women spoils that dream by reintroducing the imperative of property in their own persons.

How does any of this relate to on-line interactions, which occur not on a desert landscape but in a complex, technological society where women are supposed to command equal status with men? It accompanies us as a set of unexamined assumptions about what it means to be male or female, assumptions that we believe are rooted in the imperatives of our bodies. These assumptions follow us into the bodiless realm of cyberspace, a forum where, as one scholar put it "participants are washed clean of the stigmata of their real 'selves' and are free to invent new ones to their tastes." Perhaps some observers feel that the replication of gender roles in a context where the absence of bodies supposedly makes them superfluous proves exactly how innate those roles are. Instead, I see in the relentless attempts to interpret on-line interactions as highly gendered, an intimation of just how artificial, how created, our gender system is. If it comes "naturally," why does it need to be perpetually defended and reasserted?

Complaints about the treatment of women on line fall into three categories: that women are subjected to excessive, unwanted sexual attention, that the prevailing style of on-line discussion turns women off, and that women are singled out by male participants for exceptionally dismissive or hostile treatment. In making these assertions, the *Newsweek* article and other stories on the issue do echo grievances that some on-line women have made for years. And, without a doubt, people have encountered sexual come-ons, aggressive debating tactics, and ad hominem attacks on the Net. However, individual users interpret such events in widely different ways, and to generalize from those interpretations to describe the experiences of women and men as a whole is a rash leap indeed.

I am one of many women who don't recognize their own experience of the Net in the misogynist gauntlet described above. In researching this essay, I joined America Online and spent an hour or two "hanging out" in the real-time chat rooms reputed to be rife with sexual harassment. I received several "instant messages" from men, initiating private conversations with innocuous questions about my hometown and tenure on the service. One man politely inquired if I was interested in "hot phone talk" and just as politely bowed out when I declined. At no point did I feel harassed or treated with disrespect. If I ever want to find a phone-sex partner, I now know where to look but until then I probably won't frequent certain chat rooms.

Other women may experience a request for phone sex or even those tame instant messages as both intrusive and insulting (while still others maintain that they have received much more explicit messages and inquiries completely out of the blue). My point isn't that my reactions are the more correct, but rather that both are the reactions of women, and no journalist has any reason to believe that mine are the exception rather than the rule.

For me, the menace in sexual harassment comes from the underlying threat of rape or physical violence. I see my body as the site of my heightened vulnerability as a woman. But on line — where I have no body and neither does anyone else — I consider rape to be impossible. Not everyone agrees. Julian Dibble, in an article for *The Village Voice*, describes the repercussions of a "rape" in a multiuser dimension, or MUD, in which one user employed a subprogram called a "voodoo doll" to cause the personae of other users to perform sexual acts. Citing the "conflation of speech and act that's inevitable in any computer-mediated world," he moved toward the conclusion that "since rape can occur without any physical pain or damage, then it must be classified as a crime against the mind." Therefore, the offending user had committed something on the same "conceptual continuum" as rape.

Tellingly, the incident led to the formation of the first governmental entity on the MUD.

No doubt the cyber-rapist (who went by the nom de guerre Mr. Bungle) appreciated the elevation of his mischief-making to the rank of virtual felony: all of the outlaw glamour and none of the prison time (he was exiled from the MUD). Mr. Bungle limited his victims to personae created by women users, a choice that, in its obedience to prevailing gender roles, shaped the debate that followed his crimes. For, in accordance with the real-world understanding that women's smaller, physically weaker bodies and lower social status make them subject to violation by men, there's a troubling notion in the real and virtual worlds that women's minds are also more vulnerable to invasion, degradation, and abuse.

This sense of fragility extends beyond interactions with sexual overtones. The *Newsweek* article reports that women participants can't tolerate the harsh, contentious quality of on-line discussions, that they prefer mutual support to heated debate, and are retreating wholesale to women-only conferences and newsgroups. As someone who values on-line forums precisely because they mandate equal time for each user who chooses to take it and forestall various "alpha male" rhetorical tactics like interrupting, loudness, or exploiting the psychosocial advantages of greater size or a deeper voice, I find this perplexing and disturbing. In these laments I hear the reluctance of women to enter into the kind of robust debate that characterizes healthy public life, a willingness to let men bully us even when they've been relieved of most of their traditional advantages. Withdrawing into an electronic purdah where one will never be challenged or provoked, allowing the ludicrous ritual chest-thumping of some users to intimidate us into silence — surely women can come up with a more spirited response than this.

And of course they can, because besides being riddled with reductive stereotypes, media analyses like *Newsweek*'s simply aren't accurate. While the on-line population is predominantly male, a significant and vocal minority of women contribute regularly and more than manage to hold their own. Some of The Well's most bombastic participants are women, just as there are many tactful and conciliatory men. At least, I think there are, because, ultimately, it's impossible to be sure of anyone's biological gender on line. "Transpostites," people who pose as member of the opposite gender, are an established element of Net society, most famously a man who, pretending to be a disabled lesbian, built warm and intimate friendships with women on several CompuServe forums.

Perhaps what we should be examining is not the triumph of gender differences on the Net, but their potential blurring. In this light, *Newsweek*'s stout assertion that in cyberspace "the gender gap is real" begins to seem less objective than defensive, an insistence that on-line culture is "the same" as real life because the idea that it might be different, when it comes to gender, is too scary. If gender roles can be cast off so easily, they may be less deeply rooted, less "natural" than we believe. There may not actually be a "masculine" or "feminine" mind or outlook, but simply a conventional way of interpreting individuals that recognizes behavior seen as in accordance with their biological gender and ignores behavior that isn't.

For example, John Seabury wrote in the *New Yorker* (June 6, 1994) of his stricken reaction to his first "flame," a colorful slice of adolescent invective sent to him by an unnamed technology journalist. Reading it, he begins to "shiver" like a burn victim, an effect that worsens with repeated readings. He writes that "the technology greased the words . . . with a kind of immediacy that allowed them to slide easily into my brain." He tells his friends, his coworkers, his partner — even his mother — and, predictably, appeals to CompuServe's management for recourse — to no avail. Soon enough, he's talking about civilization and anarchy, how the liberating "lack of social barriers is also what is appalling about the net," and calling for regulation.

As a newcomer, Seabury was chided for brooding over a missive that most Net veterans would have dismissed and forgotten as the crude potshot of an envious jerk. (I can't help wondering if my fellow journalist never received hate mail in response to his other writings; this bit of e-mail seems comparable, par for the course when one assumes a public profile.) What nobody did was observe that Seabury's reaction — the shock, the feelings of violation, the appeals to his family and support network, the bootless complaints to the authorities — reads exactly like many horror stories about women's trials on the Net. Yet, because Seabury is a man, no one attributes the attack to his gender or suggests that the Net has proven an environment hostile to men. Furthermore, the idea that the Net must be more strictly governed to prevent the abuse of guys who write for the *New Yorker* seems laughable — though who's to say that Seabury's pain is less than any woman's? Who can doubt that, were he a woman, his tribulations would be seen as compelling evidence of Internet sexism?

The idea that women merit special protections in a environment as incorporeal as the Net is intimately bound up with the idea that women's minds are weak, fragile, and unsuited to the rough and tumble of public discourse. It's an argument that women should recognize with profound mistrust and resist, especially when we are used as rhetorical pawns in a battle to regulate a rare (if elite) space of gender ambiguity. When the mainstream media generalize about women's experiences on line in ways that just happen to uphold the most conventional and pernicious gender stereotypes, they can expect to be greeted with howls of disapproval from women who refuse to acquiesce in these roles and pass them on to other women.

And there are plenty of us, as The Well's response to the *Newsweek* article indicates. Women have always participated in on-line communications, women whose chosen careers in technology and the sciences have already marked them as gender-role resisters. As the schoolmarms arrive on the electronic frontier, their female predecessors find themselves cast in the role of saloon girls, their willingness to engage in "masculine" activities like verbal aggression, debate, or sexual experimentation marking them as insufficiently feminine, or "bad" women. "If that's what women on line are like, I must be a Martian," one Well woman wrote in response to the shrinking female technophobes depicted in the *Newsweek* article. Rather than relegating so many people to the status of gender aliens, we ought to reconsider how adequate those roles are to the task of describing real human beings.

*Howard Besser*

# FROM INTERNET TO INFORMATION
# SUPERHIGHWAY

In 1994 the mass media began devoting increased coverage to the
impending arrival of the information superhighway. Readers of
newspapers and popular magazines were repeatedly exposed to rosy
predictions of increased access to information, improvement of education and health care, and a diversity in home entertainment that would
all come from the promised "500 channels" of information.

While these social and recreational benefits might be a possible
result of increased channel capacity, they are certainly not the inevitable
outcome that the mass media would have us believe. Technological
developments do not in themselves provide widespread social benefits.
Both technology and social benefits are shaped by social forces that
operate on a much broader level. We need only look at similar predictions in the recent past to see that the benefits promised by a greater
channel capacity may prove to be a hollow promise.

The 1967 report of the President's Task Force on Communications
Policy made a series of recommendations on the role that should be served
by emerging cable television systems. The industry should be structured

> to cater to as wide a variety of tastes as possible, the tastes of
> small audiences and mass audiences, of cultural minorities and

of cultural majorities. Television should serve as varied as possible an array of social functions, not only entertainment and advertising . . . but also information, education, business, culture, and political expression. . . . Television should provide an effective means of local expression and local advertising, to preserve the values of localism, and to help build a sense of community. . . . [P]olicy should guard against excessive concentration in the control of communications media. (United States 1967)

A February 1973 report on the future of cable TV by the National Science Foundation was enthusiastic about what cable TV would offer:

> Public access channels available to individual citizens and community groups. . . . Churches, Boy Scouts, minority groups, high school classes, crusaders for causes — can create and show their own programs. With public access, cable can become a medium for local action instead of a distributor of prepackaged mass-consumption programs to a passive audience. New services to individual subscribers, such as televised college courses and continuing education classes in the home. Cable's capability for two-way communication between viewer and studio may in time permit doctors to participate in clinical seminars at distant hospitals, or enable viewers to register their opinions on local issues. . . . Public and private institutions might build their own two-way cable networks or lease channels to send x-rays among hospitals, exchange computer data, and hold televised conferences. (Baer 1973, 2)

They listed key features that cable would offer including: "Instruction for homebound and institutionalized persons, Preschool education, High school and post-secondary degree courses in the home, Career education and in-service training, Community information programming, Community information centers, and Municipal closed-circuit applications." (Baer 1973, 6)

The predictions made for cable television more than two decades ago sound remarkably like the predictions being made for the information superhighway today. When listening to today's predictions, we should keep in mind how empty those promises proved for cable.

## The Internet vs. the Information Superhighway

Popular discourse would have us believe that the information super-highway will just be a faster, more powerful version of the Internet. But there are key differences between these two entities, and in many ways they are based on diametrically opposed models.

### • Flat Fee vs. Pay per Use

Most Internet users are either not charged to access information or pay a low flat fee. The information superhighway, on the other hand, will likely be based upon a pay-per-use model. On a gross level, one might say that the payment model for the Internet is closer to that of broadcast (or perhaps cable) television while the model for the information superhighway is likely to be more like that of pay-per-view TV.

Flat-fee arrangements encourage exploration. Users in flat-fee environments navigate through webs of information and tend to make serendipitous discoveries. Pay-per-use environments give users the incentive to focus their attention on what they know they already want, or to look for well-known items previously recommended by others. In pay-per-use environments, people tend to follow more traditional paths of discovery and seldom explore unexpected avenues. Pay-per-use environments discourage browsing. Imagine how a person's reading habits would change if they had to pay for each article they looked at in a magazine or newspaper.

Yet many of the most interesting things we learn about or find come from following unknown routes, bumping into things we weren't looking for. And people who have to pay each time they use a piece of information are likely to increasingly rely upon specialists and experts. For example, in a situation where the reader will have to pay to read each paragraph of background on Bosnia, she is more likely to rely upon State Department summaries instead of paying to become more generally informed herself. And in the 1970s and 1980s the library world learned that the introduction of expensive pay-per-use databases discouraged individual exploration and introduced the need for intermediaries who specialized in searching techniques.

• *Privacy*

The metering that will have to accompany pay-per-view on the information superhighway will need to track everything that an individual looks at (in case she wants to challenge the bill). It will also give governmental agencies the opportunity to monitor reading habits. Many times in the past the FBI has tried to view library circulation records to see who has been reading which books. In an on-line environment, service providers can track everything a user has bought, read, or even looked at.

In an age when people engage in a wide variety of activities on line, service providers will amass a wealth of demographic and consumption information on each individual. This information will be sold to other organizations that will use it in their marketing campaigns. Some organizations are already using computers and telephone messaging systems to experiment with this kind of demographic targeting. For example, in mid-1994, *Rolling Stone* magazine announced a new telephone-based ordering system for music albums. After using previous calls to build "a profile of each caller's tastes . . . custom messages will alert them to new releases by their favorite artists or recommend artists based on previous selections" (Laura Evenson, "Phone Service Previews Albums," *San Francisco Chronicle,* June, 30, 1994). Some of the early experiments promoted as tests of interactive services on the information superhighway were actually designed to gather demographic data on users ("Interacting at the Jersey Shore: FutureVision Courts Advertisers for Bell Atlantic's Test in Toms River," *Advertising Age,* May 9, 1994).

• *Producers vs. Consumers*

On the Internet anyone can be an information provider or an information consumer. On the information superhighway most people will be relegated to the role of information consumer.

Because services like movies-on-demand will drive the technological development of the information superhighway, movies' need for high bandwidth into the home and only narrow bandwidth coming back out will likely dominate (see Howard Besser, "Movies on Demand May Significantly Change the Internet," *Bulletin of the American Association for Information Science,* October 1994). Metaphorically, this will be like a ten-lane highway coming into the home, with only a tiny path leading back out — just wide enough to take a credit card number or to answer multiple-choice questions.

This kind of asymmetrical design implies that only a limited number of sites will have the capability of outputting large volumes of bandwidth onto the information superhighway. If such a configuration becomes prevalent, this is likely to have several far-reaching results: it will inevitably lead to some form of gatekeeping. Managers of those sites will control all high-volume material that can be accessed. And for reasons of scarcity, politics, taste, or corporate preference, they will make decisions on a regular basis as to what material will be made accessible and what will not. This kind of model resembles broadcast or cable television much more than it does today's Internet.

The scarcity of outbound bandwidth will discourage individuals and small groups from becoming information producers and will further solidify their role as information consumers. "Interactivity" will be defined as responding to multiple-choice questions and entering credit card numbers on a keypad. It should come as no surprise that some of the major players trying to build the information superhighway are those who introduced televised home shopping.

• *Information vs. Entertainment*
The telecommunications industry continues to insist that functions such as entertainment and home shopping will be the driving forces behind the construction of the information superhighway. Yet there is a growing body of evidence that suggests that consumers want more information-related services, and would be more willing to pay for these than for movies-on-demand, video games, or home-shopping services.

Two surveys published in October 1994 had very similar findings. According to the *Wall Street Journal* (Bart Ziegler, "Interactive Options May Be Unwanted, Survey Indicates," October 5, 1994), a Lou Harris poll found that

> a total of 63% of consumers surveyed said they would be interested in using their TV or PC to receive health-care information, lists of government services, phone numbers of businesses and non-profit groups, product reviews and similar information. In addition, almost three-quarters said they would like to receive a

customized news report, and about half said they would like some sort of communications service, such as the ability to send messages to others. But only 40% expressed interest in movies-on-demand or in ordering sports programs, and only about a third said they want interactive shopping.

A survey commissioned by *Macworld* (Charles Piller, "Dreamnet," *Macworld*, October 1994) claims to be "one of the most extensive benchmarks of consumer demand for interactive services yet conducted"; this survey found that "consumers are much more interested in using emerging networks for information access, community involvement, self-improvement, and communication, than for entertainment." Out of a total of 26 possible on-line capabilities, respondents rated video-on-demand tenth, with only 28 percent indicating that this service was highly desirable. Much more desirable activities included on-demand access to reference materials, distance learning, interactive reports on local schools, and access to information about government services and training. Thirty-four percent of the sample was willing to pay over $10 per month for distance learning, yet only 19 percent was willing to pay that much for video-on-demand or other entertainment services.

If people say they desire informational services more than entertainment and shopping (and say that they're willing to pay for it), why does the telecommunications industry continue to focus on plans oriented toward entertainment and shopping? Because the industry believes that, in the long run, this other set of services will prove more lucrative. After all, there are numerous examples in other domains of large profits made from entertainment and shopping services but very few such examples from informational services.

It is also possible that the industry believes that popular opinion can easily be shifted from favoring informational services to favoring entertainment and shopping. For several years telecommunications industry supporters have been attempting to gain support for deregulation of that industry by citing the wealth of interesting informational services that would be available if this industry was freed from regulatory constraints. Sectors of the industry may well believe that the strength of consumer desire for the information superhighway to meet information needs (as shown in these polls) is a result of this campaign.

According to this argument, if popular opinion can be swayed in one direction, it can be shifted back in the other direction.

• *Mass Audience*

A significant amount of material placed on the Internet is designed to reach a single person, a handful of people, or a group of less than 1,000. Yet commercial distributors planning to use the information super-highway will have to reach tens (or more likely hundreds) of thousands of users merely to justify the costs of mounting multimedia servers and programs. This will inevitably result in a shifting away from the Internet's orientation towards small niche audiences; the information superhighway will be designed for a mass audience (and even niche markets will be mass markets created by joining enough small regional groups together to form a national or international mass market).

Because distributors will view their audience as a mass audience, a number of results are likely. First of all, information distributors will favor uncontroversial programs, for fear of alienating part of their audience. In recent years we have seen the extreme version of this, where controversial programs have actually been eliminated from network television and radio, cable, local broadcast stations, and even art museums due to pressure from various organizations. Perhaps less obvious is the fact that the overwhelming majority of programming focuses on elements that appeal to most people but don't offend anyone (the least common denominator); this focus is due to the orientation toward a mass audience.

For similar reasons, programs designed for mass consumption will be favored over those perceived as having a relatively narrow appeal. The result is likely to be a lack of diversity and an emphasis on whatever has mass appeal. For an example of possible long-term results from this phenomenon, we can examine the forces affecting bookstores and video stores around the country. Beginning in the late 1980s, independent book and video stores have been rapidly replaced by chain stores. Independents tend to offer a widely diverse material and in some ways use popular, mass-appeal items to subsidize more esoteric works. Chains, on the other hand, tend to carry little other than popular items. Because of the economies of scale realized by stressing mass-appeal items, chains are putting the independents out of business, and it is getting more difficult to find items that don't have mass appeal. If this carries over onto the information superhighway, we can expect that what may start out as diverse offerings will, for economic reasons, soon turn into relatively bland and homogeneous programs with mass appeal.

Other parts of the mass formula have parallels with book and video stores. Upholding a tradition they share with libraries, independent bookstores tend to take strong stands against censorship. Chain stores, on the other hand, tend to "not want to offend" and avoid carrying controversial items. (For example, after the death threat to Salman Rushdie, many independents carried both books and displays of *The Satanic Verses*, while most chains refused to even carry it.)

Chain stores tend to deal almost exclusively with major publishers and distributors who can offer them better, volume discounts and less paperwork. Independents tend to be one of the few venues for small presses or independent videos. A decade ago we saw legislative attempts to favor large studio film and video productions over independent productions in proposals to "tax" blank tapes (see, for example, "Tax on Home Videotaping Is Urged," *The New York Times,* April 22, 1982; Ernest Holsendolph, "Legislative Plan to Tax Video Recording Gear," *The New York Times,* March 12, 1982; William Raspberry, "No to the Betamax Tax," *The Washington Post,* April 29, 1983; and Lardner (1987)). While these efforts were designed to compensate major producers for illegal copying, in effect they amounted to an attempted fund transfer from independent producers (who spent a more significant percentage of their budget on blank tape) to the large studios who were to receive the "tax" distribution based upon perceived market share.

Independents tend to be close to their community and cater to particular tastes within their community. Chains, on the other hand, tend to focus on national tastes and not carry many items that may cater to primarily local or regional tastes. Some chains have been accused of trying to impose their perception of national tastes upon local communities. Blockbuster, for example, refuses to carry programs it deems controversial because of sexual or political themes, even in communities that do not find those themes offensive (see, for example, "Blockbuster to Shun Video of 'The Last Temptation,'" *The Wall Street Journal,* June 23, 1989; and "Blockbuster," *San Francisco Weekly,* March 29, 1991).

At the same time they carry movies of a violent nature even in communities that find these themes offensive. For a generation, this same mass formula allowed broadcast television to dominate the discourse over what constitutes national taste. This is likely to carry over onto the information superhighway, with video-on-demand service providers imposing their national standards upon each local community they enter. Information providers will claim that this is done for purely economic reasons — that it is not cost-justifiable to spend a fortune digitizing and mounting a program that will be of interest only to a few communities. But this amounts to censorship by economics.

## Changing Access and Relationship to Culture

As discussed above, cultural options available in an on-line environment will be dominated by mass-market productions that do not offend. But as more and more people rely on on-line access to culture, this shift is also likely to have a great effect on how people view culture, as well as on the perception and internal workings of our cultural repositories (such as museums and libraries).

As it becomes more and more convenient to view high-quality representations of cultural objects (and accompanying explanatory information) on the home computer, people are likely to visit museums less frequently. As more and more people access representations of museum objects without entering the edifice, the authority of the museum (and its personnel) will rapidly erode. In libraries, we are already beginning to see that the people who have traditionally served as caretakers of on-site collections are instead becoming designers of *access* to collections that may reside either on or off site. And as people gain the ability to seek information without the direct help of museum and library personnel, we are seeing a drastic diminution of their role as intermediaries.

As individuals look at more and more cultural objects on their workstation screens, it is likely that they will begin to confuse the representations with the original objects they represent. This is part of a general leveling effect (equating abstracts of experiences for the experiences themselves) that appears to be an integral part of contemporary life. This is not unlike viewing a video and equating that experience with watching a film in a theater — or eating at McDonald's and calling it a meal. Although in an on-line system more people gain greater access to a certain range of cultural objects, this kind of access eliminates a richness and depth of experience — what Walter Benjamin, in his classic essay "The Work of Art in the Age of Mechanical Reproduction," called the "aura" of a unique work of art.

The widespread viewing of digital images poses interesting authenticity and authorship questions. Because digital images can be seamlessly altered, how can the viewer be sure that the image on view has not been manipulated? A number of magazines have placed purposely altered images on their covers (*Time*'s June 24, 1994, darkened mug shot of O. J. Simpson, *Newsday*'s February 16, 1994, photo falsely showing Tonya Harding and Nancy Kerrigan skating together, *Spy*'s February 1993 shot of Hillary Clinton in a bondage outfit, and *Mirabella*'s September 1994 composite photo of several models' faces). Although in the above examples the magazines admitted (often in tiny print) that they altered the photos, in the future we are likely to see ever more such alterations without the publishers alerting the audience.

Having images of cultural objects widely available on line in the home is likely to lead to a proliferation of derivative art works based upon the on-line works. The ease of altering these digital images will lead individuals to make changes to them and incorporate them into larger works in a collagelike process. What we have seen with clip art and desktop publishing is likely to significantly increase as continuous-tone images of cultural objects become widely available.

When someone alters an existing image, this raises interesting questions as to who is the creator of the new work: the creator of the original work, the person who altered it, or a combination of the two? In anticipation of the widespread availability of digital works on line, copyright holders of existing images have attempted to strongly assert their intellectual property rights over works in other domains. In recent court cases Disney made R. Crumb stop using a mouse in his comic strips, a photographer won a multimillion-dollar judgment over Jeff Koons for "copying" his photograph of a man cradling a large litter of puppies, and the copyright holders of a song won a judgment against a rap group who incorporated pieces of "Pretty Woman" into one of their recordings.

Today, ownership of intellectual property rights of digital images of cultural objects is considered a great investment opportunity. The highly inflated price paid for Paramount Communications and Bill Gates's establishment of Continuum Productions to buy up electronic reproduction rights of images were early shots in what will become an economic battle over the ownership of content.

If it continues (as is likely, given the strong economic incentives in an on-line digital domain), this strong assertion of intellectual property rights will have a chilling effect upon future artistic endeavors. Postmodern art, especially, relies upon the recycling of images from the past, and the strong assertion of intellectual property rights will keep much material out of the hands of future artists. The strong assertion of intellectual property rights also has the potential of eliminating satire and will serve to limit social, political, and artistic commentary — all of which rely on being able to represent the domain that they are reacting against.

As it becomes easier and easier to obtain images and documents on line in the home, it is possible that people will download and copy these somewhat indiscriminately. The advent of the photocopy machine led researchers to become less discriminating and to copy articles of only marginal interest. This led to a glut of paper in researchers' homes and offices. Word processing led to the generation of paper drafts each time a slight change to the text was made. In a similar way, on-line access to full-text documents and digital images may lead people to accumulate items of only marginal interest. And the proliferation of images (both those available and those accumulated) may lead to a reduction in meaning and context for all of them. This leveling effect (floating in an infinite sea of images) is a likely result of information overload — we are already seeing traces of it as people are caught in the web of the Internet, not being able to discriminate between valuable and worthless information, and not seeing the context of any given piece of information.

In a way, the on-line environment of the future is the logical extension of postmodernism. As in previous incarnations (like MTV), most of our images come from the media. The images are reprocessed and recycled. In the postmodern tradition, all images (and viewpoints) have equal value; in an on-line world they're all ultimately bits and bytes. Everything is ahistorical and has no context.

It is interesting to examine the development of technological communications from generation to generation. From the radio generation to the television generation to the MTV generation to the coming virtual-reality generation, we can see a steady progression incorporating an ever quicker pace and relying on the stimulation of a larger number of senses.

One of the identifying characteristics of the information age is to get people directly to the information they need without exposing them to tangentially interesting or relevant material. Information science research in the 1980s and 1990s has focused on tailoring information to user profiles and techniques borrowed from the artificial-intelligence community in order to avoid subjecting the user to information overload. But this approach devalues serendipitous discovery.

And as this approach comes to dominate, it will help reinforce the notion (promoted by various forms of technology) that chance encounters should be avoided. From the distribution of mass-produced goods to providing a choice of movies to watch and when to watch them (rather than relying upon our local cinema), technology has always promised us more predictable, controllable experiences. In an era when most people feel pressed for time and fearful of chance encounters in a hostile world, they shun public spaces and turn to experiences that involve fewer unpredictable interactions.

Over time, our experiences with technology are replacing public spaces and human interaction. Channel-surfing on a couch in front of a 500-channel television minimizes the chance and often dangerous encounters that might take place "window shopping" past inner-city movie theaters. And computer-based "virtual" experiences (including "virtual sex") will provide us with experiences that are more predictable, less serendipitous, and less interesting than human interaction.

References

Baer, Walter S. 1973. Cable Television: A Summary Overview for Local Decision making. National Science Foundation Research Applied to National Needs Program, 134-NSF. Santa Monica: Rand (February).

Lardner, James. 1987. *Fast Forward: Hollywood, the Japanese, and the Onslaught of the VCR.* New York: Norton.

United States. 1967. President's Task Force on Communications Policy. Final Report (August 14). Washington: GPO. [Known as "the Rostow Report."]

*Jesse Drew*

# MEDIA ACTIVISM AND RADICAL DEMOCRACY

The tracks were laid not to connect internal areas one with
another, but to connect production centers with ports. The
design still resembles the fingers of an open hand: thus railroads,
so often hailed as forerunners of progress, were an impediment
to the formation and development of an internal market.
— Eduardo Galeano, *The Open Veins of Latin America*

## Progress, Profits, and Control

Popular mythology — as spread by advertising and other sources of
corporate ideology — holds that new technology is driven by the natural expansion of the market system and forward-looking entrepreneurs. In fact, this is frequently not the case. There are many instances
where the market system worked to impede, retard, or abandon new
technologies that would have benefited the public good but without
enriching the corporations. Both radio and television were held back by
patent holders who thought these new inventions were threatening the
corporate status quo. The fax machine was available in the 1930s, but
was not developed because it might have eaten into the profits of the

dominant communications companies. FM radio, which eventually opened up a tremendous amount of spectrum space, was held back for years by the influence of the major AM broadcasters. Low-power TV and low-wattage micro–FM radio transmitters have been kept off the air by archaic regulations that protect the power of for-profit broadcasters. The UHF television bands, which could have opened up an enormous number of new television channels, remain either vastly underused or marginal. Other recent examples of corporate curtailment of technology include VHS tape duplicators and DAT recorders.

Media pundits, academic futurists, and technology-industry spokespeople who uncritically promote the new technologies like to describe the upcoming technological changes as a "revolution," implying some larger social change, as if the major structural problems confronting our democracy were merely technical shortcomings. But as Ernest Mandel noted in *Late Capitalism* (1972),

> Belief in the omnipotence of technology is the specific form of bourgeois ideology in late capitalism. This ideology proclaims the ability of the existing social order gradually to eliminate all chance of crises, to find a technical solution to all its contradictions, to integrate rebellious social classes and to avoid political explosions.

The "postindustrial revolution" has not meant any great shift for the better in class, race, or gender relations: it has not made people free, happy, or equal. Rather, it has intensified the internationalization of capital and the disenfranchisement of the marginalized and the poor and has resulted in the commodification of almost every aspect of social life. It is the expansion of the commodity form into services, data, entertainment products, and information that helps obscure the true role this "software" plays in late capitalist society. (Meanwhile, the production and distribution of palpable commodities — including computers — goes on, usually "offshore," with the assembly-line work performed by a combination of sweated labor and information-rich robots, which often use techniques and processes developed by the Pentagon or with Pentagon funding.)

In a historical irony, the infrastructure created by video, cable, telephones, personal computers, and computer networks has revealed the possibility of a democratic communications system where citizens could participate equally in the reception and transmission of media and information. Technologies dedicated to the extension of government control and corporate profits turn out to have other uses: while business requires nonhierarchical, peer-to-peer communications in the era of "the virtual corporation," private citizens find that they too can take limited advantage of the same technologies. The chief obstacle for "unofficial communicators" remains that political and economic forces work against the realization of this democratic ideal.

The deregulation of industry and the destruction of antitrust laws during the Reagan-Bush years — a tendency that persists under the Clinton administration — have provided powerful catalysts for the intensive monopolization now taking place. Ben Bagdikian's estimate in the 1992 edition of *The Media Monopoly* that most of the major media are controlled by twenty-one corporations is beginning to look optimistic. When the smoke clears, we might be faced with the prospect of fewer than ten dominant megacorporations encompassing television, radio, computer networks, movie theaters, publishing houses, newspapers, and all other forms of mass media, information, and entertainment. These conglomerates will also be tied, either directly or by overlapping directorships, to the major manufacturing and financial powers.

## Strategies of Containment

The history of wireless telegraphy/telephony — "radio" — demonstrates the way in which prevailing conditions shape the development and use of technical inventions. In the early post–World War I period, RCA and other interested corporations — GE, AT&T, Westinghouse — thought of radio as parallel to the telephone, another way of transmitting business information. But as a message service it had one very serious disadvantage: there was no privacy. There were many, however, who saw this liability as its unique asset; namely, a novel and cheap method of "broadcasting" publicity. Dealers, department stores, hotels, banks, restaurants, and churches mounted transmitters on their roofs.

Those who saw radio as essentially a commercial medium were soon challenged by arguments on behalf of public-interest broadcasting. As eloquently laid out by R. W. McChesney in his study of the labor movement and its involvement in the early days of radio ("Labor and the Marketplace of Ideas: WCFL and the Battle for Labor Radio Broadcasting, 1927–1934," *Journalism Monographs* (1992)), it was not a foregone conclusion that radio would benefit exclusively the large corporate interests.

The establishment of the FCC and the writing into law of the Communications Act of 1934 represented a total victory for communications corporations and a defeat for the proponents of educational broadcasting and the public good. One company alone, RCA, benefited from the 1934 communications act by gaining two networks, major manufacturing facilities of radio equipment, international ship-to-shore communications systems, and the majority of clear-channel radio stations. Other major beneficiaries were AT&T, General Electric, and Westinghouse. With the control over the nation's communications system firmly in the hands of these corporations, the nation's airwaves quickly became, in the words of Senator Burton K. Wheeler, a "pawnshop." Thus the promise of radio, with its two-way potential to reach a mass audience simultaneously and promote a diversity of viewpoints was essentially dead, diverted into a system that sought to push consumer products to the largest audience possible. Even Lee de Forest, the man credited with the invention of the vacuum tube that helped launch radio broadcasting, condemned the system of commercial broadcasting. He wrote to the National Association of Broadcasters (*Chicago Tribune*, October 28, 1946) complaining of what the industry had done to his "child": "You have made of him a laughing stock to intelligence . . . you have cut time into tiny segments called spots (more rightly stains) herewith the occasional fine program is periodically smeared with impudent insistence to buy and try."

The sordid history of the selling-out of radio was recapitulated with television, which was immediately developed as a tool serving business interests and promoting the views of the dominant political class, rather than for the interests of a diverse public. Language fought for by public-interest groups like labor unions, community organizations, and civil rights organizations, was included in the legal framework of television broadcasting, but it did little to ameliorate the crass commercial nature of the industry. Public television, under the thumb of the Washington political machine and at the mercy of corporate funding, has not provided a convincing alternative.

## Activist Theory

> In its present form, equipment like television or film does not serve communication but prevents it. It allows no reciprocal action between transmitter and receiver; technically speaking, it reduces feedback to the lowest point compatible with the system.
> — Hans Magnus Enzensberger, *The Consciousness Industry*

The continuing "debate" over the National Information Infrastructure, the so-called information superhighway, has so far been a staged event. Mass-circulation magazines like *Newsweek* and *Time* have run cover stories about the coming "revolution" that are little more than regurgitations of industry PR. Lifestyle magazines like *Wired* and *Mondo 2000* stimulate consumer demand for new gadgets and informed acquiescence to governmental and corporate policies — in the name of a spurious "liberation" and "empowerment." Buzzwords like "interactive" and "choice" are frequently employed. "Choice," however, means no more than a consumer choosing a product offered by the seller of services or commodities, perhaps video-on-demand or database services. "Interactive" means the ability to punch in your credit card number and order products via screen commands from home. The very language of interactivity and the new data communications is being subverted by this emphasis on one-way flow and sales.

The mass media usually limit the critical response to this celebration of new communication technology to academics who accuse the new electronic media of eroding a past, higher civilization of print-based culture. Much of the writing of this group calls for curtailing the new technologies. In an earlier time, their counterparts lamented the rise of the paperback book and its "cheapening of culture."

What is lacking, between the hype and the jeremiads, is a perspective on the new media that combines criticism of nefarious uses (commercial, governmental, military) with an understanding of the democratic possibilities, in order to develop a theory of communications technologies as tools for social progress. A handful of theorists have advanced the idea that new technologies can be used for progressive ends. In the 1930s writers such as Walter Benjamin and Bertolt Brecht began to lay such a foundation. This unofficial current has continued in the works of Hans Magnus Enzensberger, Alexander

Kluge, Raymond Williams, Dee Dee Halleck, and Douglas Kellner, among others.

## Media Activism in TV

Along with new technological developments and in reaction to the conservative agenda of the media monopolies, a new breed of political activists has emerged to fight the influence of corporate control over mass media and to construct new forms of alternative media. These media activists contend that new communications technologies can be used to benefit large sectors of society historically underserved and left without a voice. Media activists are often low-budget producers engaged in grass-roots political activism, such as camcorder documentary makers, community radio DJs, 'zine publishers, and computer hackers. While theoretical understanding of the social implications of low-budget media remains underdeveloped, media activists have successfully demonstrated the usefulness of, for example, the video camcorder as a means of democratic expression and organization. But while media activists and the general public can use consumer equipment for their productions, they are severely hampered by the corporate stranglehold over the means of exhibition and distribution.

The most dramatic consequences of camcorder populism are exemplified by George Holiday's Rodney King footage, which helped touch off the L.A. rebellion. Camcorders have been used as electronic witnesses to assist against injustices and abuses in Bosnia, China, Romania, the Amazon rain forest, Native American lands, Palestine, Haiti, and Tibet. In Northern Iraq, Kurdistan guerrillas have built their own television system around camcorder technology in defiance of Iraqi repression. In 1992, Witness sent out forty Hi-8 camcorders, fifty Polaroid cameras, and a hundred fax machines to human rights organizations around the world to assist in this electronic battle against injustice.

In the United States, the availability of inexpensive camcorders has spawned numerous organizations and collectives over the years, such as Alternative Views (Austin), Labor Beat (Chicago), Not Channel Zero (New York), Paper Tiger TV (New York and San Francisco), Collision Course (San Francisco and Buffalo), DIVA (video arm of

ACT-UP), Flying Focus Video Collective (Portland, Oregon), the 8mm News Collective (Buffalo), UPPNET (labor video).

Paper Tiger Television is often referred to as a pioneer of "guerrilla television," having produced programs since 1981. As a member of the Paper Tiger Collective since 1984, I have personally experienced the victories as well as the frustrations of trying to harness video technology as a tool for social change.

Paper Tiger TV took advantage of the creation of public-access television stations, which activists demanded of cable companies in exchange for the right to lay cable in cities and towns. Early PTTV productions were usually produced live-to-tape in public-access studios and featured activists and critics giving a critical "reading" of a particular media publication. Early shows featured people like Herb Schiller on the *New York Times,* Murray Bookchin on *Time,* Donna Haraway on *National Geographic,* and Alexander Cockburn on the *Washington Post.* These shows tore at the veil of "professionalism" and "objectivity" that often cloak the propagandistic function of these organs of power and cultural control. While deconstructing the messages of these media, often in a humorous fashion, the shows also made links between their corporate ownership and their vested interest in maintaining the status quo. By being cablecast on public-access stations in different cities, and by screenings in public sites such as storefronts and classrooms, these programs have helped to renew interest in media criticism and media "literacy."

In the late 1980s the focus of the group shifted as the camera and the video deck converged into the camcorder, which allowed guerrilla television producers to go out into the community as never before. The dropping price of the camcorder made it easier to prove the point that you don't have to be a network or a TV station to make your own media. A case in point is *Drawing the Line at Pittston* (1990), a tape made in collaboration between Paper Tiger members and striking mine workers in Virginia's Pittston mines. Footage captured by PTTV members and the miners showed a story that had been largely ignored by the mainstream media. Besides being shown on public-access stations, this tape was used to raise support from other unions in other states. The videotape was also used to document that violence on the picket lines was being instigated by the police and state troopers. When I showed this tape at a large meeting of union members, the response was over-whelming. For many, it was the first time they had ever seen television depict the struggle of working people in a positive and realistic light.

In search of better means of distribution, Paper Tiger TV discovered that the lowered cost of satellite technology allowed them to broadcast rather than just dub and mail their tapes to other public-access stations. The proliferation of satellites and the lowered cost of renting transponder time allows activists to uplink and beam alternative programming side-by-side with corporate data flows and HBO. Thus was born Deep Dish Television, which has used satellite transmission to uplink many hours of programming produced by hundreds of media activists and guerrilla television producers. Rather than just use Deep Dish to further the reach of Paper Tiger, the philosophy has been to rely on local producers and activists to supply material with national interest and impact. For example, in a program on housing or health care, rather than produce a single national program by professionals, Deep Dish would call for tapes from local activists and mediamakers. The best of these tapes would be integrated into a single program that Deep Dish would uplink. This approach allows activists around the country to see what others are doing, while letting the local activists who created the tape see their activities in a national framework. Deep Dish has beamed up hundreds of hours of programs on topics such as racism, Central America, labor issues, women, the environment, and children. The satellite "footprint" allows for the programs to be picked up by hundreds of public-access stations as well as anyone (mostly rural people) with a backyard dish.

## Deep Dish and the Gulf War

The need for alternative television communications was driven home by the launching of the murderous assault by the U.S. on the people of Iraq in 1992. In an attempt to reassert U.S. control over Middle Eastern oil fields and divert domestic problems into flag-waving patriotic hysteria, George Bush launched Operation Desert Storm. In a display of submission and one-sidedness that would have made Joseph Goebbels envious, the U.S. corporate media stifled alternative expression and acted as the official mouthpiece of the government.

Paper Tiger TV had created The Gulf Crisis TV Project months before the war began in an effort to show an alternative view of the developing situation in the Gulf. These tapes included interviews with army deserters, antiwar veterans, and analysts like Noam Chomsky, Edward Said, and Daniel Ellsberg, and showed how different communities were organizing to resist the move to war.

By the time the first four half-hour tapes were broadcast on Deep Dish TV, war fever had already swept the country. The interest in these tapes was enormous, as evidenced by the number of requests that poured into the Gulf Crisis office in New York and the Paper Tiger TV offices in New York and San Francisco. Work began on six more half-hour programs that documented the growing antiwar movement and included information we felt important enough to be passed around the unofficial and official censorship of the U.S. media.

These tapes were shown to many large audiences in theaters, universities, and public spaces, which had the effect of breaking the spiral of silence that the U.S. media was promulgating in refusing to depict any mention of domestic dissent. The tapes were heavily used by activists all across the United States to help build local antiwar activities. Thousands of copies were dubbed and redubbed and passed between friends, relatives and coworkers.

Pressure put on the Public Broadcast System to show an alternative view allowed the Gulf Crisis TV Project to be shown in major cities such as New York and Los Angeles. When the San Francisco PBS station refused to program the show, we held a public screening on the wall of their building by using a video projector and a portable public-address system. In San Francisco, where there had been hundreds of thousands of demonstrators and more than one thousand arrests for civil disobedience, our local Paper Tiger group started a weekly cable series as a tool for antiwar activists. We had dozens of video volunteers shooting antiwar activities on a daily basis, from blockades to teach-ins. We would end each show with announcements of all the events for the coming week. And we were able to use some of this footage as defense evidence for people arrested in police sweeps. Gulf Crisis TV material was also shown on national televisions systems in Europe, Asia, and elsewhere.

Occasionally there will be extreme situations like the Gulf War in which the advantages of alternative electronic media will leap to the fore. Of course, in comparison to the monolithic media structures erected by the Time-Warner-Viacom-TCI-GE nexus, the faint snarling of Paper Tiger TV can seem ineffective. Groups like Paper Tiger TV lack the resources for building mass audiences through traditional means such as advertising and slick production values. Furthermore, in attempting to reach an invisible audience "out there," we lack the resources for even gauging our effectiveness. However, I believe that groups like Paper Tiger, while not close to threatening mainstream television, have developed important followings on the margins of society. Public-access and alternative viewing sites have played important roles in maintaining and cultivating oppositional and minority cultures. Public-access stations, for example, frequently have a high concentration of programs by African Americans, gay and lesbian groups, Latinos, radical groups, foreign-language-speaking groups, and other communities ignored by mainstream media.

## Activist Radio and Beyond

Alongside the development of the camcorder, advances in semiconductors and integrated circuits have made it possible to assemble sophisticated FM radio transmitters that fit in the palm of the hand. To avoid detection, an entire radio station can be placed in a shoe box and carted from house to house. These circuits are tunable to unused frequencies on the FM dial and are very stable so as to avoid interference with adjacent channels. The combination of large listening audiences and the ease of transmitter assembly and operation make this form of communication attractive to many grass-roots organizations and activists.

Grass-roots radio stations are found across the country, from San Francisco to Phoenix to the Blue Ridge Mountains to Manhattan's Lower East Side. Perhaps the most famous of these stations is Black Liberation Radio, run by African-American activist M'banna Kantako in Springfield, Illinois. His station, operating with only a few watts of power in the heart of a central housing project, reaches most of the city's black population. Of course, Black Liberation Radio — like Free Radio Berkeley and San Francisco Liberation Radio — has faced severe legal reprisals by the FCC and relentless police harassment.

Consumer electronics has supplied other tools that can be used for democratic expansion of the social arena. Desktop publishing software has enabled the production of thousands of new 'zines. Micro-radio transmitters are being used from Eastern Europe to Chiapas. Homemade television transmitters have already gone on the air in San Francisco, Prague, and Vancouver. The Internet allows U.S. users to exchange information with Cuba, circumventing the U.S. blockade. Native Americans of the Western Shoshone Nation use fax machines, police scanners, and radiotelephones to defend themselves against recent attacks by the federal government.

## Confrontation in Cyberspace

The explosive growth of the Internet and the evolution of proposals for a National Information Infrastructure have heightened the struggle between the democratic and the corporate communications models. Ironies abound: the Internet began as a Department of Defense experiment in constructing a decentralized computer network that could survive a nuclear war; but this network outgrew the experiment, to become a relatively cheap and open way of communicating between computer users. Now its low cost and ease of access have begun to seem like liabilities to the profit-seekers who dream of privatizing and "improving" the Internet to the point of charging for every mouse click.

But then much of the hard work of laying down the initial "trails" that led to the information superhighway was done by outcast hackers, computer hobbyists, and activists who created the bulletin board services and other refuges in cyberspace for freethinking people. Many of these computer bulletin boards served as transit points around which the current Internet system grew. Indeed, the foundations of micro-computing itself can be attributed to this community of hackers, whose members thought they saw in the microcomputer a more democratic and decentralized response to the monolithic nightmare of IBM main-frames and Orwellian information databases.

The computer subculture spawned by the microcomputer created a hacker's code of ethics, based on the ideas of "shareware" and freedom to access information, ideas antithetical to the kind of commercial development now planned by corporate entrepreneurs. Thus the early slogan of hacking: "Information wants to be free." There are hundreds if not thousands of bulletin board services, Freenets, and other computer sites that, along with such established progressive sites as PeaceNet and The Well, make up an informal alternative information structure.

## The Prospects for Democratic Media

Whatever the origins of the technologies employed, the fundamental questions about design and use remain. Will the information superhighway be decentralized, inexpensive, and open? Will it facilitate grassroots production and distribution? Or will it permit the media giants to establish a one-way flow of home shopping and movies-on-demand?

Engineering decisions are rarely based on public need and improving our quality of life; they are usually made with considerations of profit in mind. Only seldom does the public get involved in technological decisions, and this is normally in a reactive fashion when people are trying to stop something that has already been put into use, such as nuclear power and pesticides.

Critical engineering decisions being made right now will have dramatic repercussions on whether communications will help to democratize collective life. The nature of the coming information superhighway will powerfully affect our future, yet discussion and planning for it take place mostly behind closed doors. Decisions about privacy on line — of great concern to many citizens who want their computer communications to be secure — are made behind a veil of secrecy.

Certainly these are not decisions best made by specialists, politicians, and industrialists. Rather, as Raymond Williams points out in *Resources of Hope* (1989), society

> has to get rid of the idea that communication is the business of a minority talking to, instructing, leading on, the majority. It has, finally, to get rid of the false ideology of communications as we have received it: the ideology of people who are interested in communications only as a way of controlling people, or of making money out of them.

In many ways we are at a greater disadvantage than we were at the beginnings of radio, because modern corporations and the communications conglomerates are much more powerful, while the voice of the public is much weaker and less well organized. Traditional opposition — trade unions, immigrant and minority organizations, and others — is very weak. Our culture is dominated by the influence of the corporate mentality as never before. Our future depends not only on the specific form that new technologies take, but on what kind of social and political structure we create and to what ends this society uses these technologies. A popular movement for social change must take advantage of the new technologies to further democratize the nation and to empower the disenfranchised. It is not the technology that will revolutionize society, but a movement of millions that must transform society.

*Richard E. Sclove*

# MAKING TECHNOLOGY DEMOCRATIC

A century and a half ago Alexis de Tocqueville described a politically exuberant United States in which steaming locomotives could not restrain citizens' ceaseless involvement in politics and community life:

> In some countries the inhabitants seem unwilling to avail themselves of the political privileges which the law gives them; it would seem that they set too high a value upon their time to spend it on the interests of the community; and they shut themselves up in a narrow selfishness. . . . But if an American were condemned to confine his activity to his own affairs, he would be robbed of one half of his existence; he would feel an immense void in the life which he is accustomed to lead, and his wretchedness would be unbearable.

That is not today's United States, where a bare majority of eligible voters participate in presidential elections, while engagement in local politics is sometimes much less — despite the fantastic growth in the means of communication. The causes of political disengagement are complex, but I believe that technology has played a more significant and intricate part than is commonly credited. Consider an instructive story from across the Atlantic:

During the early 1970s running water was installed in the houses of Ibieca, a small village in northeastern Spain. With pipes running directly to their homes, Ibiecans no longer had to fetch water from the village fountain. As families gradually purchased washing machines, fewer women gathered to scrub laundry by hand at the village washbasin. Arduous tasks were rendered technologically superfluous, but village social life unexpectedly changed. The public fountain and washbasin, once scenes of vigorous social interaction, became nearly deserted. Men began losing their sense of easy familiarity with the children and donkeys that formerly helped them haul water. Women stopped gathering at the washbasin to intermix scrubbing with politically empowering gossip about men and village life. In hindsight this emerges as a crucial step in a broader process through which Ibiecans came to relinquish the strong bonds — with one another, animals, and the land — that had knit them into a community (Harding 1984). Painful in itself, such loss of community carries a specific political risk as well: as social ties weaken, so does a people's capacity to mobilize for political action (Bowles and Gintis 1986).

Like Ibiecans, we acquiesce in seemingly benign or innocuous technological changes. Ibiecans opted for technological innovations promising convenience, productivity, and economic growth. But they didn't reckon on the hidden costs: deepening inequality, social alienation, and community dissolution and political disempowerment.

### Economic Performance vs. Democracy

In recent years American progressives have generally agreed with conservatives on orienting technology policy toward conventional economic objectives: productivity, growth, and international competitiveness. But where conservative rhetoric favors laissez-faire means toward these ends, proposed progressive technology policies endorse government intervention, including, for instance, new civilian authorities for guiding technological development away from misguided militarism toward basic societal needs such as health care, environmental protection, infrastructural improvements, and competitive consumer products. The more daring of these proposals are avowedly populist, seeking decentralized institutions for grass-roots empowerment within scientific and technological decision making. Nevertheless, insofar as progressives concur on orienting technology policy fundamentally toward economic objectives, their disagreement with conservatives involves means more than ends. Progressive strategies proposed to date will consistently miss their mark until they come more fully to terms with technologies' latent role in shaping fundamental social and political relations.

Contemporary technologies contribute indirectly to diverse social ills, and in particular conspire in subtle ways to significantly hinder participatory democratic decision making. Yet if technologies' social and political potency is not taken into account, the best we can hope for is a variant of Ibieca's fate: improvements in productivity or in addressing basic social needs that are nonetheless associated with further unintended declines in political engagement, attenuation of community bonds, experiential divorce from nature, individual purposelessness, and expanding disparities in wealth. Is that what we dream and sweat for?

Fortunately, it is possible to envision alternative technological strategies and designs that, while still fulfilling vital economic and social needs, can also help sustain democratic community, civic engagement, and social justice. Thus the point is not to reject all technology outright — clearly a ludicrous proposition — but rather to become more discriminating in how we design, choose, and use technologies. Moving in this direction would entail broadening progressive technological agendas to comprise a complete democratic politics of technology: democratic institutions for evolving democratic technologies. It would also entail taking the political risk of publicly granting equal or greater precedence, within our strategies, to advancing democracy over fulfilling short-run economic objectives.

While I think the latter political gamble justified, I can respect the judgment of those who differ — provided their choice is an informed one. Their decision to champion a social needs-driven technology policy, with emphasis upon improved economic productivity and growth, is politically safer and perhaps on that basis warranted. But to date there is no evidence that this strategy's proponents have considered the offsetting risk that, even if "successful," it will deliver us into a world of deepening inequality, passivity, and alienation. In contrast, the decision to espouse a democratic politics of technology, while confronting serious short-run political obstacles, embodies the chance to move into a world of expanding personal and collective empowerment, a world in which concocting and consuming "new and improved!" deodorants, hi-tech designer sneakers, or chip-laden electronic gadgetry need no longer preempt the quest to revitalize democracy, address real social needs, and fulfill other high human aspirations.

This doesn't mean that technologies are necessarily the single most important factor influencing political life. Rather, technologies are sufficiently important — and so inextricably intertwined with other factors, such as legislation, the distribution of wealth, race and gender relations, international affairs, and so on, that we must learn to subject technologies to the same rigorous political scrutiny and involvement that should be accorded to those other factors. But can I really be serious about attending to democracy before economics? Absolutely. After all, isn't one lesson of the collapse of Eastern European Communism that an economy unsupervised by democracy is not only bad because it is undemocratic, but furthermore risks injustice, ecological spoliation, and gross economic inefficiency? Or suppose that the United States' founders had decided to subject each article of the proposed Bill of Rights to an economic cost-benefit test? Would we be satisfied if they had rejected protection of civil liberties on the grounds that the supposed benefits could not hold a candle against the obvious economic costs?

Technologies don't merely deliver sundry consumer benefits (not to mention sundry hazards and irritations); they constitute part of a society's core political infrastructure. Technologies do this by establishing an intricate and pervasive network of structurally consequential social influences, opportunities, constraints, and inducements. Notwithstanding our obliviousness, this network is as significant as, say, the more explicit network of opportunities, constraints, and inducements established by a modern society's legal and tax codes, together with their enforcement mechanisms. Until we act on this insight, technological developments — whether guided by the existing market and government policies or by commonly espoused progressive alternatives — will ultimately frustrate democracy, as well as the other social objectives that a vigorous democracy would choose to pursue.

Technological democratization — democratic processes for generating a more democratic technological order — is thus vital for several basic reasons: First, it is a crucial but almost entirely neglected requirement for advancing societal democratization generally. Second, it directly addresses a number of commonly perceived social problems, such as the decline of face-to-face community and the degraded nature of work. Third, by contributing to overall societal democratization, technological democratization would contribute to the broad-based fairness and empowerment needed to begin effectively addressing technologies' other social consequences (i.e., those not specifically political) as well as chronic societal problems otherwise not attributable to technology.

### Technology as Social Structure

A crucial step toward grasping the need for a democratic politics of technology thus involves learning to see technologies as more than mere tools for accomplishing narrowly defined objectives. Technologies also represent an important kind of social structure. By "social structure" I mean the background features that help define and regulate social life. Familiar examples include laws, dominant political and economic institutions (such as legislatures, courts, and corporations), and systems of cultural belief. All of these — like technologies — qualify as social structures by virtue of being social creations that profoundly influence one another's evolution, as well as the course of history and texture of daily life. In so doing, social structures shape and help constitute a society's fundamental political relationships and processes. To appreciate the subtle means through which technologies can exert structural influence, reconsider Ibieca.

First, upsetting Ibieca's traditional pattern of water use compromised important means through which the village perpetuated itself as a self-conscious community. Thus technologies indeed help structure social relations. But notice that technologies tend to do this independently of their nominally intended (or "focal") purposes. We do not normally regard fountains, pipes (or, for that matter, microwave ovens, hypodermic syringes, garden hoses, or numerically controlled machine tools) as devices that shape patterns of human relationship, but that is nevertheless one of their pervasive latent (or "nonfocal") tendencies.

Second, clusters of focally unrelated technologies often interact to produce structural results that no one technology would produce alone. In Ibieca, introducing water pipes added incentive for also replacing donkeys with tractors in field work. (The fewer tasks a donkey is asked to perform, the less economical it is to maintain it.) This eliminated any remaining practical use for donkeys, while increasing villagers' dependence on outside jobs for the cash needed to finance and operate their new tractors and washing machines. Thus it's not enough to consider just one kind of technology at a time (water pipes); we must analyze all the different artifacts, practices, and systems that jointly comprise a society's entire technological order (water pipes, washing machines, and tractors as an interdependent system).

•

The fact that any given technology often produces its most profound social effects indirectly — and generally in concert with other, seemingly unrelated technologies — suggests a debilitating weakness in technological strategies focused solely on addressing social needs. Of course, we should attempt to address such needs. But we must do so through a strategy able to ensure that diverse technologies' combined focal and nonfocal results contribute constructively to the grass-roots empowerment and other social objectives we hope to advance. If Ibiecans want new ways of managing their water use or plowing their fields, can they be helped to craft means that won't entail indirectly sacrificing other desirable aspects of community life? (Perhaps a centrally located laundry cooperative instead of individual household washing machines?) Or, if Americans are prepared to begin substituting better health care or public transportation for our self-destructive addiction to advanced military hardware, can we be sure not to adopt approaches that will inadvertently reinforce national social malignancies, such as the decay of inner cities and the corresponding rise of atomistic suburbs?

The preceding technological insights also highlight the inadequacy in becoming too preoccupied either with sexy, cutting-edge technologies — such as biotechnologies or "the information superhighway" — or with technologies that are focally political — such as computerized voting, public affairs television, or weaponry (see Richard Sclove and Jeffrey Scheuer, "The Ghost in the Modem," the *Washington Post,* May 29, 1994). While it is foolish to overlook such technologies, it is just as wrong to ignore the more familiar or focally innocuous technologies — the pipes, washing machines, air conditioners, electricity distribution networks, and so on — whose combined significance is typically at least as important.

## Technology and Democracy

My approach to devising a technological strategy attuned to the preceding insights is grounded in the belief that people should be able to help shape the basic social circumstances of their lives. This implies struggling to organize societies along relatively equal and participatory lines, a vision of egalitarian decentralization and confederation that Benjamin Barber (1984) labels "strong democracy." Historic examples exhibiting aspects of this ideal include New England town meetings, the confederation of self-governing Swiss villages and cantons, and the tradition of trial by jury. Strong democracy is apparent also in the methods or aspirations of various social movements, such as the late-nineteenth-century American Farmers Alliance, the 1960s U.S. civil rights movement, and the 1980s Polish Solidarity movement. In each of these cases, ordinary people claimed the rights and responsibilities of active citizens concerned with basic social issues.

This procedural standard of strong democracy is complemented by a substantive standard: in their political involvements citizens ought to grant precedence to perpetuating their society's basic character as a strong democracy. Apart from this substantive obligation, we should be free to attend as we wish to our various other personal and shared concerns.

This model of democracy, even in schematic form, is sufficient for deriving a simple but compelling theory of democracy and technology: if citizens ought to be empowered to participate in determining their society's basic structure, and technologies are an important species of social structure, it follows that technological design and practice should be democratized. From strong democracy's complementary substantive and procedural standards, we can see that this involves two components: Substantively, technologies must become compatible with our fundamental interest in strong democracy itself. Procedurally, we require expanded opportunities for people from all walks of life to participate in shaping their technological order.

### Design Criteria for Democratic Technologies

The following outline presents some criteria for distinguishing among technologies based on their structural compatibility with democracy. I characterize these criteria as "provisional" because this list is neither complete nor definitive. Rather, I hope to provoke political discussion that can gradually issue in a broadened and improved set of criteria. (The complete derivation and justification for these criteria is given in my forthcoming book from Guilford Press.)

*A Provisional System of Design Criteria for Democratic Technologies*

  • *Toward democratic community —*
A. Seek balance between communitarian/cooperative, individualized, and intercommunity technologies. Avoid technologies that establish authoritarian social relationships.

  • *Toward democratic work —*
B. Seek a diverse array of flexibly schedulable, self-actualizing technological practices. Avoid meaningless, debilitating, or otherwise autonomy-impairing technological practices.

  • *Toward democratic politics —*
C. Seek technologies that can help enable disadvantaged individuals and groups to participate fully in social and political life. Avoid technologies that support illegitimately hierarchical power relations between groups, organizations, or polities.

  • *To help secure democratic self-governance —*
D. Restrict the distribution of potentially adverse consequences (e.g., environmental or social harms) to within the boundaries of local political jurisdictions.
E. Seek relative local economic self-reliance. Avoid technologies that promote dependency and loss of local autonomy.
F. Seek technologies (including an architecture of public space)

compatible with globally aware, egalitarian political decentralization and federation.

• *To help perpetuate democratic social structures —*
G. Seek ecological sustainability. Avoid technologies that are ecologically destructive of human health, survival, and the perpetuation of democratic institutions.
H. Seek local technological flexibility and global technological pluralism.

As the outline indicates, each of these criteria is intended to have some direct and important bearing upon one of strong democracy's three general institutional requirements: democratic community, democratic work, or democratic politics (For examples illustrating the practical significance of these criteria, see Sclove 1993). In making technological decisions, there are of course many other issues that we might want to address. But we should attend first and especially to democracy, because it enables us to freely and fairly decide what other considerations to take into account in our technological (and nontechnological) decision making. Until we do this, current technologies will continue to hinder the advancement of other social objectives.

## Democratic Design Criteria vs. Progressive Proposals

The preceding discussion suggests an important deficiency in the familiar progressive call to "rebuild America's crumbling technological infrastructure." Insofar as that infrastructure needs repair or modernization, shouldn't we be striving to rebuild an infrastructure more amenable to local democratic governance? For instance, there are ways of reducing and efficiently managing industrial and municipal waste, and conserving or producing energy, that can be deployed and administered with extensive local involvement. Neighborhood-scale seasonal storage of solar energy for heating homes and buildings, as pioneered in Scandinavia, is one example. Indeed, unless localities regain more control of their own sustaining infrastructures, there will be diminished incentive for the grass-roots political involvement essential to strong democracy, because the majority of important decisions affecting localities will continue to be made elsewhere.

Another important feature of all the preceding criteria is that they are designed to work together as a complementary system applied to an entire technological order, not only to single technologies. This, too, can remedy weaknesses prevalent among progressive technology policies. For instance, advocates of workplace democracy aim admirably to fulfill criteria A and B (cooperation and creativity), but generally fail to inquire whether the resulting goods and services are socially benign. But if we competitively and democratically produce democratically dubious technologies — say, chemical weapons, or certain consumer electronics (walkmen and the like) that can erode social interaction and solidarity — our achievement will remain ambiguous.

Similarly, in an era of explosive popular concern with acid rain, atmospheric ozone depletion, and global greenhouse warming, few doubt the necessity of devising more ecologically sustainable technological activities (criterion G). Yet environmentalists sometimes imagine that, as a social basis for technological design, sustainability alone is sufficient. The incompleteness here becomes apparent if one recalls sewage system configurations that have both protected public health and helped subvert local democratic governance, or thinks of Singapore's relatively stringent environmental policies coupled with a starkly authoritarian political regime (Stan Sesser, "A Reporter at Large: A Nation of Contradictions," *The New Yorker*, January 13, 1992). In evaluating technology, we must learn to take into account all technologies, their focal and nonfocal effects, and the manifold respects in which technologies influence political relations.

### Toward a Democratic Politics of Technology

Democratic design criteria are essential to democratic politics of technology, but only if coupled with institutions for greater popular involvement in all domains of technological decision making. This suggests the need to establish extensive opportunities for popular participation in contesting and applying democratic design criteria (in communities, workplaces, and all other social realms), setting research and development (R&D) priorities, and governing important technological systems.

Thus the basic idea would be to open, democratize, and partly decentralize pertinent government agencies, create avenues for worker and community involvement in corporate R&D and strategic planning, and generally strengthen societal capabilities to monitor and, as needed, guide technologies' cumulative political and social consequences (For specific suggestions, see Sclove 1994). We also need political strategies and target policies for accomplishing these objectives, preferably — in the interest of political practicability — building outward from popular movements and technological initiatives that already exist. The end of the Cold War makes the time ripe for such efforts.

There are many prototypes for institutions or processes supporting greater popular involvement in technological decisions. For instance, the Community Health Decisions (CHD) movement has developed grass-roots processes for forging popular consensus on ethical principles governing health care policy and technology. One of the movement's accomplishments is to have organized dozens of community meetings throughout Oregon where social values were articulated, debated, and prioritized in ways that appreciably influenced bold reforms in the state's health care system (Jennings et al. 1990). The CHD movement could provide one model for future forums in which citizens would debate more general democratic design criteria.

In several states, coalitions of peace activists, labor unions, business leaders, community groups, and government officials have successfully launched broadly democratic processes for diversifying regional economies away from military production toward areas of civilian need. For example, responding to grass-roots pressure, Washington State has established a broadly representative citizen advisory group that monitors the state's military economy; assesses post–Cold War economic adjustment problems; identifies nonmilitary economic needs, opportunities, and capabilities; and promulgates action plans to help military-dependent communities diversify prior to defense-related job cutbacks (Cassidy 1992). One can interpret this, among other ways, as an excellent example of learning to use social criteria to evaluate and redirect an entire existing technological order.

Over the past twenty years the Dutch have developed a network of public "science shops," supported by nearby university staff and students, where citizen groups receive free assistance in learning to address social issues that have technical components. One science shop helped a local environmental group document heavy metals contamination in vegetables, inducing the Dutch government to sponsor a major clean-up, as well as process modifications in the responsible polluting metal-working plants. Holland's science shops have proven so successful a vehicle for citizen empowerment within technological decision making that they have prompted related efforts in other European nations (Shulman 1988).

Traditional Amish communities, often misperceived as technologically naive or backward, have pioneered popular deliberative processes for screening technologies based on their cumulative social effects, in effect attending to many of the criteria outlined above. One method — certainly a worthy candidate for emulation — is to place the adoption of certain new technologies under one-year probation, in order to discover empirically what the social effects will be. For instance, Amish dairy communities in east central Illinois ran a one-year trial with diesel-powered bulk milk tanks before judging them socially acceptable, whereas other Amish communities have used the results of similar voluntary social trials to decide to prohibit household telephones or personal computers. The preceding examples are atypical in the extent to which they deviate from the norm of expert decisions, bureaucratic machinations, or unregulated market interactions that determine most technology choices. But for that very reason they provide crucial evidence that, given the right institutional circumstances, lay citizens can make reasonable technological decisions reflecting democratic priorities that otherwise lie fallow.

### Participatory Research, Development, and Design

Democratic processes for technological choice and governance are vital, yet still hardly worth the effort unless participants have a sufficiently broad range of alternative technologies among which to choose. Hence it is essential to press further, weaving democracy into the very fabric of technological innovation: the research, development, and design (RD&D) processes. Consider some examples:

## Democratic Design of Workplace Technology

In Scandinavia unionized newspaper graphics workers — in collaboration with sympathetic university technical researchers, a Swedish government laboratory, and a state-owned publishing company — succeeded during the early 1980s in inventing a form of computer software unique in its day. Instead of following emerging trends toward routinized or mechanized newspaper layout, this software foreshadowed some of the capabilities later embodied in desktop publishing programs, thus enabling printers and graphic artists to continue to exercise considerable creativity in page design (Martin 1987). Known as UTOPIA, this project is an instance of broadened participation in the RD&D process leading to a design innovation that, in turn, supports one condition of democracy: creative work (see criterion B, above).

UTOPIA was less ambitious in scope than several other attempts at participatory design within the workplace. (For instance, in the 1970s workers at Britain's Lucas Aerospace Corporation sought not only to democratize their own work processes, but also to produce more socially useful products (Wainwright and Elliott 1982)). But UTOPIA went further practically, moving beyond developing prototypes to a successfully marketed design innovation. This instance of collaboration between workers and technical experts — initially limited to a single technology within a single industry — occurred under unusually favorable social and political conditions. Sweden's work force is 85 percent unionized, and the nation's pro-labor Social Democratic party had held power during most of the past half century, providing a supportive legislative context.

## Participatory Architecture

Compared with the relative paucity of attempted exercises in participatory design of machinery, appliances, or technical infrastructure, there is a rich history of citizen participation in architectural design. The range of stories is extremely diverse, including cases in which it proved difficult to motivate participation or in which the design outcome was little different from what architects working alone might have devised. But in other cases highly novel designs resulted — sometimes evoking resistance from powerful social interests and institutions.

One example is the "Zone Sociale" at the Catholic University of Louvain Medical School in Brussels. Here in 1969 students insisted that new university housing mitigate the alienating experience of the adjacent massive hospital architecture. Architect Lucien Kroll established an open-ended, participatory design process that elicited complex organic forms, richly diverse patterns of social interaction — e.g., a nursery school situated near administrative offices and a bar — and a complex network of pedestrian paths, gardens, and public spaces. Walls and floors of dwellings were movable, so that students could design their own living spaces. Construction workers were offered design principles and constraints rather than finalized blueprints, and encouraged to generate sculptural forms. They did, and some returned on Sundays with their families to show off their art. The project's structural engineers were initially baffled by the level of spontaneity and playfulness but they were gradually reeducated into competent participants. All appears to have proceeded splendidly for some years, until the university administration became alarmed at its loss of control over the process. With students away on vacation, they fired Kroll and halted further construction (Kroll 1984).

### Feminist Design

What would happen if women played a greater part in RD&D? One indication originates from feminists, who have long been critical of housing designs and urban layouts that enforce social isolation and arduous, unpaid domestic labor upon women (Hayden 1984). As ever more households deviate from the conventional norm of mother and children supported financially by a working husband, such criticism has sharpened.

One historical response has been to suggest that if women were more actively engaged in design, they might promote more shared neighborhood facilities (such as day care, laundries, or kitchens) and closer co-location of homes, workplaces, and commercial and public facilities. Realized examples exist in London, Stockholm, and Providence, Rhode Island.

Another approach has been pioneered by artist and former overworked-mother Frances GABe, who devoted several decades to inventing a self-cleaning house. "In GABe's house, dishes are washed in the cupboard, clothes are cleaned in the closets, and the rest of the house sparkles after a humid misting and blow dry!" (Zimmerman 1986).

Still other feminists have established women's computer networks, or seek alternatives to proletarianized female office work, to urban transportation networks that are insensitive to women's needs, and to new reproductive technologies that divorce women from control of their bodies. One explicit feminist complaint against current reproductive technologies — such as various infertility treatments, surrogate mothering, hysterectomy, and abortion — is that women have played little part in guiding medical R&D agendas, the results of which now impose upon women (that is, more than men) agonizing moral dilemmas that might otherwise have been averted or structured very differently.

### Barrier-Free Design on the Information Superhighway

During the past two decades, there has been substantial innovation in designing "barrier-free" equipment, buildings, and public spaces responsive to the needs of people with physical disabilities. Much of the impetus comes from disabled citizens themselves, who have played roles ranging from mobilizing to redefine and assert their needs, to participating in inventing or evaluating design solutions. For example, prototypes of the Kurzweil Reading Machine — which uses computer voice-synthesis to read aloud typed text — were tested by over 150 blind users. In an eighteenth-month period these users made over 100 recommendations, many of which were incorporated into later versions of the device. This is a prime example of how technology related to the National Information Infrastructure could be researched and designed — a far cry from the current efforts at designing technologies to create a market for, say, video-on-demand and home shopping.

Of course, there are large-scale design decisions to be made regarding the technical architecture of the NII; it is of paramount importance that these decisions be arrived at democratically because these "technical" considerations could well build in biases affecting the usability of the communications network for, on the one hand, corporate media conglomerates and, on the other, popular institutions and private individuals. It is also essential that planning for the NII take into consideration the needs of users and nonusers alike. Just as the creation of the interstate highway system had profound effects on everyone's life — reorganizing commerce and the geography of cities, contributing to the destruction of old neighborhoods and the ruining of public transportation, facilitating the growth of suburbs — the coming rewiring of the United States will doubtless change life for everyone, including the multitude who, by circumstance or choice, never use computers.

•

These and other examples of participatory design demonstrate convincingly that it is realistic to envision a much wider range of people participating in technological research, development, and design. Moreover, the resulting designs have indeed tended to establish a broadened and substantively more democratic menu of subsequent technological choices. But we also learn that democratic politics of technology is unlikely to arrive without a struggle: many participatory design exercises encounter fierce opposition from powerful institutions — opposition engendered not because the exercises are failing, but because they are succeeding.

Still, advocates of participatory RD&D have often elected to state their case entirely in terms of nonparticipants' material interests. They have also cited the contribution that participation can make to improved productivity or to more satisfactory design solutions. These can be reasonable arguments, and sometimes effective ones. But they neglect the argument that the opportunity to participate in RD&D is a matter of moral right — essential to individual moral autonomy, to human dignity, to democratic self-governance, and to generating technologies more compatible with democracy.

References

Barber, Benjamin. 1984. *Strong Democracy: Participatory Politics for a New Age.* Berkeley: University of California Press.

Bowles, Samuel, and Gintis, Herbert. 1986. *Democracy and Capitalism: Property, Community, and the Contradictions of Modern Social Thought.* New York: Basic Books.

Cassidy, Kevin S. 1992. "Defense Conversion: Economic Planning and Democratic Participation." *Science, Technology, and Human Values* 17:3 (Summer).

Harding, Susan Friend. 1984. *Remaking Ibieca: Rural Life in Aragon under Franco.* Chapel Hill: University of North Carolina Press.

Hayden, Dolores. 1984. *Redesigning the American Dream: The Future of Housing, Work, and Family Life.* New York: W. W. Norton.

Jennings, Bruce, et al. 1990. "Grassroots Bioethics Revisited: Health Care Priorities and Community Values." *Hastings Center Report* 20:5 (September-October).

Kroll, Lucien. 1984. "Anarchitecture." In *The Scope of Social Architecture.* Edited by Richard C. Hatch. New York: Van Nostrand Reinhold.

Martin, Andrew. 1987. "Unions, the Quality of Work, and Technological Change in Sweden." In *Worker Participation and the Politics of Reform.* Edited by Carmen Sirianni. Philadelphia: Temple University Press.

Sclove, Richard E. 1993. "Technological Politics as if Democracy Really Mattered: Choices Confronting Progressives." In *Technology for the Common Good.* Washington: Institute for Policy Studies.

———. 1994. "Democratizing Technology." *Chronicle of Higher Education* (January 12).

Shulman, Seth. 1988. "Mr. Wizard's Wetenschapswinkel." *Technology Review* 91:5 (July).

Wainwright, Hilary, and Elliott, Dave. 1982. *The Lucas Plan: A New Trade Unionism in the Making?* London: Allison and Busby.

Zimmerman, Jan. 1986. *Once Upon the Future: A Woman's Guide to Tomorrow's Technology.* New York: Pandora.

REWIRING THE BODY

*Kevin Robins and Les Levidow*

# SOLDIER, CYBORG, CITIZEN

Virtual culture raises questions about who and what we are. It encourages us to see ourselves as if we were cybernetic organisms — confusing the technological and organic, the inner and outer realms, simulation and reality, freedom and control. For the most part, the virtual life has been understood in terms of the possibilities it opens up; in terms of a new age of Promethean adventure. As the real world is seen to be giving way to the virtual, there are those who have a growing sense of freedom and empowerment. They believe that we shall assume new powers of creativity and imagination; that virtual technologies will help us to transcend mundane reality and "to dwell empowered or enlightened on other mythic planes" (Benedikt 1991, 6). The figure of the cyborg — a hybrid of the human and the technological — has become the (mythological) focus of these "postmodern" desires and fantasies.

Idealism and utopianism are fundamental aspects of our technological culture. We are easily seduced by technological promises. It will be hard to resist the virtual life. But of course we also know that the (Enlightenment) dream has always been haunted by its monster. The cyborg struggles to turn a blind eye to what has been technologically denied and repressed in human nature.

The cyborg self can be characterized as follows: through a paranoid rationality, expressed in the machine-like self, we combine an omnipotent phantasy of self-control with fear and aggression directed against the emotional and bodily limitations of mere mortals. Through regression to a phantasy of infantile omnipotence, we deny our dependency upon nature, upon our own nature, upon the "bloody mess" of organic nature. We phantasize about controlling the world, freezing historical forces, and, if necessary, even destroying them in rage; we thereby contain our anxiety in the name of maintaining rational control (Levidow and Robins 1989, 172).

Vision and image technologies mediate the construction of the cyborg self. The so-called Gulf War highlighted their role. In a very real sense, the screen became the scene of the war: the military encountered its enemy targets in the form of electronic images. The world of simulation somehow screened out the catastrophic dimension of the real and murderous attack.

As the Gulf War also brought home to us, it was not just military personnel who became caught up in this technological psychosis. The "Nintendo war" involved and implicated home audiences, who took pleasure in watching the official images of war, often compulsively so. How was it possible to achieve this popular engagement? How were viewers locked into the war through their TV screens? How is the cyborg self generalized to the society at large?

### The Military Cyborg

War converts fear and anxiety into perceptions of external threat; it then mobilizes defenses against alien and thing-like enemies. In this process, new image and vision technologies can play a central role. Combat is increasingly mediated through the computer screen. Combatants are involved in a kind of remotely exhilarating tele-action — tele-present and tele-engaged in the theater of war, sanitized of its bloody reality. Killing is done "at a distance," through technological mediation, without the shock of direct confrontation. The victims become psychologically invisible. The soldier appears to achieve a moral dissociation; the targeted "things" on the screen do not seem to implicate him in a moral relationship.

Moreover, by fetishizing electronic "information" for its precision and omniscience, military force comes to imagine itself in terms of the mechanical or cybernetic qualities that are designed into computers. The operator behaves as a virtual cyborg in the real-time, man-machine interface that structures military weapons systems. A new "cyborg soldier" is constructed and programmed to fit integrally into weapons systems. By training for endurance, the soldier attempts to overcome biological limits, to better respond to real-time "information" about enemy movements. By disciplining his "mindware" and acting on the world through computer simulations, the soldier can remain all the more removed from the bloody consequences of his actions (Gray 1989).

In the Gulf War, the cyborg soldier was complemented by new "smart" weapons. Although the view from a B-52 bomb bay already distanced the attacker from any human victims, new weapons rationalized military vision even further. Paradoxically, the Gulf video images gave us closer visual proximity between weapon and target, but at the same time greater psychological distance. The missile-nose view of the target simulated a superreal closeness that no human being could ever attain. This remote-intimate viewing extended the moral detachment that characterized earlier military technologies (Robins and Levidow 1991, 325).

It was the ultimate voyeurism: to see the target hit from the vantage point of the weapon. An inhuman perspective. Yet this kind of watching could sustain the moral detachment of earlier military technologies. Seeing was split off from feeling; the visible was separated from the sense of pain and death. Through the long lens the enemy remained a faceless alien. Her and his bodily existence was derealized (Robins and Levidow 1991). Military attack took the form of thinglike relations between people and social relations between things, as if destroying inanimate objects. Perversely, war appeared as it was (Levidow 1994).

In targeting and monitoring the attack, a real-time simulation depended upon prior surveillance of the enemy, conceptualized as a "target-rich environment." In the five months preceding the January 1991 attack on Iraq, the U.S. war machine devoted laborious "software work" to mapping and plotting strategic installations there. The concept of "legitimate military target" extended from military bases and the presidential palace, to major highways, factories, water supplies, and power stations. The basic means of survival for an entire population were reduced to "targeting information." Enemy threats — real or imaginary, human or machine — became precise grid locations, abstracted from their human context.

This computer simulation prepared and encouraged an omnipotence phantasy, a phantasy of total control over things. At the same time, the phantasized omnipotence required the containment of anxieties about impotence and vulnerability. The drive for electronic omniscience both evoked and contained anxiety about unseen threats. Designed to prepare real-time attacks, an electronic panopticon intensified the paranoiac features of earlier omnipotence phantasies. Through these technological attempts at ordering a disorderly world, uncertainty was rendered intolerable.

Any attempt to evade penetration by the West's high-tech panopticon simply confirmed the guilt and irrationality of the devious Arab enemy. Any optical evasion became an omnipresent, unseen threat of the unknown that must be exterminated. This paranoid logic complemented the U.S.A.'s tendency to abandon the Cold War rationales for its electronic surveillance and weaponry, now being redesigned explicitly for attacking the Third World (Klare 1991).

In the Gulf episode, the U.S. military portrayed the Iraqi forces as in hiding. When Saddam decided to avoid a direct military confrontation with the U.S. coalition's air force, he was described as "hunkering down," almost cheating the surveillance systems of the West's rational game plan (Levidow and Robins 1991). Iraq's caution was personified as the backward Arab playing the coy virgin: "Saddam's armies last week seemed to be enacting a travesty of the Arab motif of veiling and concealment ... [Saddam] makes a fairly gaudy display of mystique" (*Time*, February 4, 1991). Such language updated an earlier cultural stereotyping of the mysterious Orient (Said 1985).

The racist logic emerged more clearly after the U.S. massacre of civilians in the Amariya air-raid shelter. In this case, unusually, TV pictures showed us hundreds of shrouded corpses. In response, the U.S. authorities insisted that they had recorded a precise hit on a "positively identified military target"; they even blamed Saddam for putting civilians in the bunker (Kellner 1992, 297–309). The U.S. continued to cite its surgical precision as moral legitimation — even though it was the precise targeting that allowed the missile to enter the ventilation shaft and incinerate all the people inside the shelter.

## Constructing the Viewer

This combined logic of fear and aggression is not just a military phenomenon. The Gulf War showed how much we, the home viewers of the Nintendo war, were also implicated in the logic of fear, paranoia, and aggression. As seen on network TV, the video-game images were crucial in recruiting support for the U.S.–led attack.

The images evoked an audience familiarity with video games, thus offering a vicarious real-time participation. Video games in the wider culture are also about the mastery of anxiety and the mobilization of omnipotence phantasies; these psychic dimensions correspond to the cyborg logic of the military "game." The parallel with weapons systems runs deep; after all, some innovators have alternated between designing military and entertainment versions of interactive simulation technology.

Where the Gulf massacre publicly enacted phantasies, video games privatize them. The processes of anxiety and control are actively structured by the computer-video microworld, with its compulsive task of achieving "perfect mastery" (Levidow and Robins 1989, 172–75). In particular, the video game is a psychodynamic process of projecting and managing internal threats: "The actual performance required of us in the video game is like being permanently connected to broadcast television's exciting live event." Video games elicit young boys' phantasies of exploring the damage done inside the mother's body; here the male "fears both his own destructiveness and a fantasied retaliation from the object of his destructive fantasies." (Skirrow 1986, 121–22)

Video games can thus be understood as a paranoiac environment that induces a sense of paranoia by dissolving any distinction between the doer and the viewer. Driven by the structure of the video game, the player is constantly defending himself, or the entire universe, from destructive forces. The play becomes a compulsive, pleasurable repetition of a life-and-death performance. Yet the player's anxiety can never be finally mastered by that vicariously dangerous play. He engages in a characteristic repetition, often described as "video-game addiction" (129–33).

While the video game simulates a real-time event, the Gulf episode took such images as its reality. The Gulf War was "total television," an entertainment form that merged military and media planning

(Engelhardt 1992). "The Pentagon, and its corporate suppliers, became the producers and the sponsors of the sounds and images, while the 'news' became a form of military advertising" (Stam 1992, 112). How, then, did this infotainment engage its audience, even construct the viewer?

The home audience, which seemed to take great pleasure in its viewing, was also implicated in collective phantasies. Primitive anxieties were evoked and structured by a pervasive cultural rhetoric, which gave specific meanings to the electronic images. The Iraqi state, even an entire society, was personified as an irrational monster, "a new Hitler," from whom we must be saved. The sadistic "rape of Kuwait" posed a threat of symbolic buggery against the West, even a threat to civilization itself. Saddam seemed to personify a sadistic, unpredictable, limitless violence. He was a "madman" who transgressed the combined rules of morality and rationality. By exaggerating claims about Iraq's nuclear weapons development and speculating on its chemical weapons, the mass media portrayed a regional aggressor as a global threat of annihilation.

In the face of perceived threats, viewers were infantilized, leading them to welcome a strong savior who was apparently wielding a civilized violence on behalf of international law. When the West's attack transparently went beyond the official mission to "Free Kuwait," the ensuing destruction resonated with popular wishes to remove the source of primitive anxiety — only by civilized means, of course. In contrast to Saddam's sadistic Scuddish violence, the West was imagined to be inflicting a morally based violence; "our" missiles acted as exterminating angels, by virtue of their precision and rationality. This good/evil split permitted Western TV audiences to deny the barbarism within their own civilization, to deny the internal sources of its violence, and to treat its destructive hatred as an enemy threat (Aksoy and Robins 1991). Western rationality became inseparable from a paranoid projection that conflated, and confused, internal and external threats.

Bombed facilities were rhetorically personalized as Saddam's military machine. In this way, "the media turned Iraq into one vast faceless extension of their demonized leader" (Stam 1992, 114). The mass media also adopted U.S. military euphemisms that further reified the massacre. This language denied the human qualities of the victims, while attributing such qualities to inanimate objects: for example, "smart" bombs "killed" Iraqi equipment. Home viewers could thereby detach themselves morally from the human consequences of surgical strikes against the evil, nonhuman forces personified by Saddam (Kellner 1992, 247).

The images did far more than sanitize death. The video-game war also combined viewer and doer: "telespectators were made to see from the bomber's perspective" (Stam 1992, 104; cf. Levidow and Robins 1991). With missile cameras "the sectors of destruction and information became almost completely synonymous" (Wark 1991, 15). The images involved us as vicarious participants in destroying perceived threats to our bodily integrity, our physical existence, and our social order. Indeed, we could feel a pleasurable identification with high-tech violence against a barbaric enemy (Broughton 1993).

In this paranoid rationality, the problem is less about people accepting the literal "truth" of propaganda images than about seeking refuge from anxiety. "The danger is that people will choose fantasy, and fantasy identification with power, over a threatening or intolerably dislocating social reality" (Rosler 1991, 63). As an antiwar poster warned, "You are the Target Audience": "When you watch the news, you are invited to enter into a pact. You are expected to believe in the same system, share the same values and goals."

Even those disposed to be critical could find certain parts of themselves consenting to this pact. The war images both evoked and contained primitive anxieties in all of us. We were confronted with invasive and induced feelings, and found ourselves experiencing feelings and thinking thoughts that were in an important sense not entirely our own. We had to reckon with feeling-states that seemed to inhabit and impose themselves on us, irrespective of our conscious desires. As Baudrillard (1991, 12–13) puts it, "we were all held hostage by the intoxication of the media."

## The Social Cyborg

The Gulf massacre brought home to us the role of high-tech systems in mass psychopathology. This episode belied the naive hopes of those who have idealized electronic information — as an instrument of participatory democracy, as a social prosthesis, or even as inherent resistance to the commodity form. Rather, electronic systems constituted a paranoiac environment; mediating an omnipotence phantasy, they converted internal threats into thing-like enemies, symbolizing rage at our bodily limitations.

In the paranoid-schizoid mode, "the self is predominantly a self as object, a self that is buffeted by thoughts, feelings, and perceptions as if they were external forces or physical objects occupying or bombarding oneself" (Ogden 1989, 21–22). The fear and pain that was imminent within ourselves is evacuated and experienced as the danger imminent within the other. Once they are externalized in this way, "the establishment can get to work, offering its protection, keeping the threat at bay, zapping intruders, policing the boundaries" (Hoggett 1992, 346).

Screen, surveillance, and simulation technologies have become fundamental to this "protection." It now involves new global networks of sensors keeping track of worldwide targets, in real time. Vision technologies appear to enhance security through continuous monitoring of the globe as a danger-rich environment. With the spread of such defense systems, "real enemies" become elusive and omnipresent. The technologies now monitor an unidentified and amorphous threat "out there," both within and beyond the boundaries of the Western world.

For example, Britain has an estimated 200,000 video surveillance cameras, many of them continuously monitoring main streets and shopping centers. It is argued that the cameras not only help police to identify criminals, but also make people feel safer. If so, then the cameras compensate for — and even intensify — the social isolation that makes us feels vulnerable in the first place. This form of "security" further constructs our lives as social relations between things, just as a Nintendo war "protected" us from any human relation with its victims.

The Other is an unseen and invisible threat, detectable only through electronic surveillance and mediation. The technological systems generate a structural paranoia: their panoptic vigilance requires the existence of a virtual enemy. It is not only the state that is caught up in this logic of paranoid rationality. Its psychic defense, which underlay the 1980s Star Wars project, became a mass-culture recruitment drive during the Gulf War. As in that "virtual war," now the whole society is caught up in "this visualization of things, this hypervisibility, this hyperpredictability and programming, hyperprogramming, of things" (Baudrillard 1993, 251).

Through electronic mediation, this aspect of war pervades wider areas of our lives, thus socializing the cyborg self. We fear ourselves and each other, while identifying with an omnipotence phantasy of technological power. The question remains: instead of infantilizing us, can electronic mediation help us to handle our fears and to identify with fellow targets of the paranoid panopticon?

### References

Aksoy, A., and K. Robins. 1991. "Exterminating Angels: Morality, Technology and Violence in the Gulf War." *Science as Culture* 12.

Baudrillard, J. 1991. *La Guerre du Golfe n'a pas eu lieu.* Paris: Editions Galilée.

———. 1993. "Hyperreal America." *Economy and Society* 22 (2).

Benedikt, M. 1991. "Introduction." In *Cyberspace: First Steps.* Edited by M. Benedikt. Cambridge, Mass.: MIT Press.

Broughton, J. 1993. "The Pleasures of the Gulf War." In *Recent Trends in Theoretical Psychology III.* Edited by R. Stam. New York: Springer.

Engelhardt, T. 1992. "The Gulf War as Total Television." *The Nation* (May 11).

Gray, C. H. 1989. "The Cyborg Soldier." In *Cyborg Worlds: The Military Information Society.* Edited by L. Levidow and K. Robins. London: Free Association/New York: Columbia University.

Hoggett, P. 1992. "A Place for Experience: a Psychoanalytic Perspective on Boundary, Identity, and Culture." *Environment and Planning D: Society and Space* 10 (3).

Kellner, D. 1992. *The Persian Gulf TV War.* Boulder: Westview.

Klare, M. 1991. "Behind Desert Storm: The New Military Paradigm." *Technology Review* (May-June).

Levidow, L. 1994. "The Gulf Massacre as Paranoid Rationality." In *Ideologies of Technology.* Edited by T. Druckrey and G. Bender. Seattle: Bay Press.

Levidow, L., and K., Robins, eds. 1989. *Cyborg Worlds: The Military Information Society.* London: Free Association/New York: Columbia University.

Levidow, L., and K. Robins. 1991. "Vision Wars." *Race and Class* 32(4).

Ogden, T. 1989. *The Primitive Edge of Experience.* Northvale, New Jersey: Jason Aronson.

Robins, K., and L. Levidow. 1991. "The Eye of the Storm." *Screen* 32(3).

Rosler, M. 1991. "Image Simulation, Computer Manipulations." *Ten 8* 2(2).

Said, E. 1985. *Orientalism.* Harmondsworth: Penguin.

Skirrow, G. 1986. "Hellivision: An Analysis of Video Games." In *High Theory, Low Culture.* Edited by C. MacCabe. Manchester: Manchester University Press.

Stam, R. 1992. "Mobilising Fictions: The Gulf War, the Media, and the Recruitment of the Spectator." *Public Culture* 4(2).

Wark, M. 1991. "War TV in the Gulf." *Meanjin* 50(1).

# BODY, BRAIN, AND COMMUNICATION

*George, I understand you want to make a disclaimer about computers before we begin?*

Yes. I simply want to say that I am not a computer curmudgeon. Whatever I say today has nothing to do with feeling that the clock ought to be turned back, that computers are terrible things for mankind or anything of that sort. I work on a computer, I love it. I communicate by e-mail, and it is very important that I do so. I do research with people who design computational models of mind. I have the greatest respect for them as colleagues and for their work and I think there are enormous and quite obvious advantages in computer technology that are for the better. So, with that disclaimer, let me talk about things that perhaps are mistaken or oversold.

*Perhaps we could start this way: You are well known for your work on language and metaphor and in particular for a criticism of the conduit metaphor in relation to language. Can you tell us what the conduit metaphor is? And why are you critical of it? And how the conduit metaphor relates to computers?*

The conduit metaphor is a basic metaphor that was discovered by Michael Reddy. He observed that our major metaphor for communication comes out of a general metaphor for the mind in which ideas are taken as objects and thought is taken as the manipulation of objects. An important part of that metaphor is that memory is "storage." Hence when you store something in memory you either have to retrieve it or get it to come to you, you recall it. As Reddy observed, communication in that metaphor is the following: ideas are objects that you can put into words, so that language is seen as a container for ideas, and you send ideas in words over a conduit, a channel of communication to someone else who then extracts the ideas from the words.

Reddy shows that this is the major metaphor that we have for communication and he gives lots of examples: "I got that idea *across* to him" or "Did you *get* what I was saying?" or "It *went right over* my head" or "You try to *pack* too many ideas into too few words." A great many expressions are based on the conduit metaphor. One of them is that the meaning is right there *in* the words.

*What is implied by this view of language as communication by conduit?*

One entailment of the conduit metaphor is that the meaning, the ideas, can be extracted and can exist independently of people. Moreover, that in communication, when communication occurs, what happens is that somebody extracts the same object, the same idea, from the language that the speaker put into it. So the conduit metaphor suggests that meaning is a thing and that the hearer pulls out the same meaning from the words and that it can exist independently of beings who understand words.

*That probably does seems like an attractive idea to a telephone engineer. It seems to describe quite well what is going on.*

You are bringing up the question of information theory — the whole understanding of information theory in the popular domain as opposed to information theory as a technical subject, which has to do with signals. Information theory as a popular idea is very much like the conduit metaphor. This, as Reddy points out, is the most common view of what communication and information are. And theories of teaching are based on it. When you say, "We are going to stuff this into your mind" and "You have got to regurgitate it on the exam," and so on, you are talking about the conduit metaphor, and in this view of teaching what the teacher tries to communicate to the students is actually communicated to them.

Now that is an attractive idea, and there are a set of cases where it seems to work. For example: We are now drinking tea. If I say to you, there is tea in my cup. There is no reason to think that you would have any problem understanding what a cup is, what tea is, and what it means for tea to be in the cup. The conduit metaphor works pretty well as a way of understanding what is involved in that communication. But there are a lot of cases where it just fails; in fact, it fails in most cases.

For example, in order for the conduit metaphor to work, the speaker and the hearer must be speaking the same language. If I speak to someone in English, who doesn't know English, obviously it isn't going to work. Not only must people be speaking the same language, but they have to have the same conceptual system. They have to be able to conceptualize things in the same way. So if I speak to another speaker of English, from a very different subculture, about a subject where the difference in subcultures matters a great deal, then we may not be communicating. My ideas will not be "extracted" from my words.

The other person I am talking to has to be able to have the right conceptual system to be able to understand what it is I am saying — to make anything like the same sense out of it. In addition, the person I am talking to may have to have pretty much the same kinds of relevant life experiences; he must understand the context in pretty much the same way. If someone understands the context in a totally different way, then the conduit metaphor fails. There is no lack of ways in which the conduit metaphor fails. The conduit metaphor says if you put your ideas in the right words, communication should just work. But communication isn't so simple. Communication is difficult and it takes a lot of effort. What the conduit metaphor does is hide all the effort involved in communication.

The view of information as something that is separable from human beings is an entailment of the conduit metaphor. It seems natural because that is our major metaphor for communication. Most people don't even see it as a metaphor; they see it as just a definition of communication. As a result, again as Reddy points out, one of the consequences of this is that people think that information is in books. If ideas can be put into words and words are in books, then the ideas can be in books, and the books can be in the libraries — or the ideas can be coded into the computer and therefore the information can be in the computer.

*How is that wrong?*

It is wrong in the following way: Let's suppose that we have books on ancient Greek philosophy. Let's suppose we stop training people to speak ancient Greek. Suppose nowhere in the world can people speak ancient Greek and suppose no one learns ancient Greek philosophy anymore. Can you just go to those books in ancient Greek, about ancient Greek philosophy, and understand them? Clearly, the answer is no. So there is no information in the books per se.

You have to have people who understand the language, who understand the historical context, who understand the ideas involved and the conceptual systems involved. The same thing is true of "information" in the computer. In order for anybody to understand "information," they have to put an interpretation on what comes out of the machine. This is a major problem for all software designers. It is not news to anyone who actually designs software, because the problem for software designers is that people are likely to misinterpret what is intended by the designer. "User-friendly" software is software that is likely to be understood by the person using it. Information is not straightforwardly in the computer — you have to have human beings trained to understand things in a certain very specific way before it makes any sense to talk about having "information" in a computer. What are the consequences of that? Well, there are a great many. For example, take the claim that we now have more information at our fingertips than ever before.

*A very common claim. But it's true, isn't it?*

It is not clear that it is true. Let's take an example: On the World Wide Web there is a lot of software that I have available to me that I could put on my computer. But I don't know how to use most of that software. That software is not all information *for me*. It might *become* information for me, if I were to learn certain things, but right now it isn't information for me.

Now, let's take another kind of case. One of the awful things about the conduit metaphor is that it assumes that meaning is objective. So, for example, let's take a clear case where meaning need not be objective. Suppose you consider the FBI files. They are encoded on computers. There are all kinds of data put on those files that is collected by agents, and these files have been collected over the past forty or fifty years. For all I know there might be a file on me! I would doubt that what an FBI agent wrote about me twenty-five years ago is objective. What goes into the FBI computer is not information in any neutral sense. It is something that has been subject to interpretation and upon being seen in a different context can be interpreted in a different way.

Thus it is not obvious that the FBI's computer has a lot of "information" about some particular person on whom it has a large file. It has what somebody has put in the computer given what they understood and what they took that to mean. But that is not objective "information" about that person. Does the FBI computer contain objective "information" about people? It may very well not. The FBI files are an extreme case. If you want to take an even more extreme case, look at the KGB files. Do you trust what the KGB has in its files? Do they have a lot of "information" about Americans in their files? It is a very funny idea to think that they have "information" about us given what has been put in, under what circumstances, and for what purposes.

You go from there to information on your credit file. There is "information" in your credit file about when you did and didn't pay your credit card on time and things of this sort. That in some way is "objective" information, but of course there are circumstances, interpretations, and so on because that information is used for a purpose. It is used for a purpose of deciding whether you should get a loan or get credit — it has to do with whether you are trustworthy. That is not an objective matter. Your trustworthiness is not information that can be in a computer. The only information that can be in the computer is whether a certain bill got paid on time, and things of that sort.

*Now, I take it that this is always going to be a problem if language is ever reduced to writing. Are you suggesting that it is now acutely more of a problem, given the recent advances in the technics of communication and information?*

That is exactly right. It is. Of course, it was already a problem with writing. But it is more of a problem when you have artificial intelligence programs taking databases and then reconfiguring them, interpreting them in other ways, making computations based on them. These so-called "intelligent" programs aren't intelligent. The programs just follow algorithms that someone made up. And a conclusion can be arrived at on the basis of such an algorithm. An algorithm might be applied to your credit-rating file to decide whether you should get a loan. The algorithm doesn't know you and cannot decide if you are trustworthy. Such algorithms are being used to make decisions about your life on the basis of the kind of so-called "information" in some computer.

*How did the epithet "intelligent" ever get to be applied to algorithms in a computer?*

That is a long and interesting story. The first part of the story has to do with formal logic. Gottlob Frege and Bertrand Russell were the developers of mathematical logic. Russell claimed that human rationality could be characterized by mathematical logic. Now, mathematical logic is the precursor to computer programs. The computer database is based on what is called a model for predicate calculus. It has a bunch of entities with properties and relations. All the standard models for first-order logic look like that. And the way in which symbols are manipulated in a computer program comes out of the same kind of mathematics that was developed for the theory of proofs — sometimes also called "recursive function theory," sometimes "the theory of formal systems" — but it's all the same form of mathematics. The idea was that if humans reasoned using mathematical logic, then a computer could reproduce that form of reasoning. If mathematical logic could characterize what human intelligence was, the computer could be intelligent.

That's what lies behind that idea. It's false, an utterly false notion, but a lot of people believe it, a lot of people still think it's true. There are several things behind this that are metaphorical. We saw in the conduit metaphor that people don't realize that the conduit metaphor is a metaphor; similarly, a lot of people don't realize that the metaphor "thought is mathematical logic" is a metaphor. It isn't true; it is very far from being accurate. So it is important to understand that, one, it is a metaphor, and, two, that it is false in a great many ways.

*Do you say it is false because you know it is some other way?*

Yes, we know a number of reasons why it is false. The first reason is that it is based on the assumption that reason is disembodied, that reason can be separated from the body and the brain, that it can be characterized in terms of pure form. This is an idea that goes back at least to Descartes.

What has been discovered in the cognitive sciences in the last fifteen or twenty years is that reason is embodied, that concepts are embodied — they have to do with how we function in the world, how we perceive things, how our brains are organized, and so on. It is not a matter of disembodied computation. Moreover the mechanisms of reason have turned out to be not at all just mathematical logic. There are many other very different mechanisms at work.

Humans think in terms of what are called "image schemas" — these are schematic spatial relations. For example, if you take the concept "in," it is based on what is called the "container schema," a bounded region of space. The concepts "from" and "to" are based on a "source-path-goal schema," and so on. Different languages organize these schemas in different ways. The schemas are embodied; they are not just disembodied symbols. They have topological and orientational properties that have to do with the way bodies are organized. Mathematical logic just does not capture all of this.

Secondly, there is a lot of reasoning that is metaphorical. As we saw, the conduit metaphor is part of a larger metaphor for understanding what thought is. In general, the way we understand thought is through a set of metaphors. These metaphors are not characterizable in mathematical logic. They *do* have entailments but not of the kind that logicians have talked about. For example, take classical categories as defined within mathematical logic; namely, by a list of necessary and sufficient conditions. For the most part, human beings don't think in terms of such categories. Humans think in terms of categories that have very different properties: they may be graded (or fuzzy), they may be radial (having central members and extending to other noncentral members), they may have a "prototype" structure, where you reason in terms of typical cases, ideal cases, stereotypes of a social nature, and so on. In short, most of the actual reasoning that humans do is not characterizable by mathematical logic.

*Does anything follow, then, vis-à-vis modern technologies of communica-*
*tion, from the central fact that human reasoning is embodied in the ways*
*you describe? Is it grounds for relating face-to-face, for "keeping it oral"?*
*Isn't it an argument against certain kinds of mediation, against virtuality?*

There is indeed a lot that follows for face-to-face communication —
and I don't mean face-to-face communication over a video screen. I
mean where there is a body present, where there is body language being
shown, where there is emotion being shown. For instance, in a book I
just received today, *Descartes' Error*, Antonio Damasio, a neuroscientist
who works with patients who have brain injuries, discusses the case of
a man who has all his rational faculties — he can reason abstractly quite
well — but has lost the capacity to feel. He can feel nothing about poetry, music, sex. It turns out that he does very badly in reasoning about
his own life. His life is a mess. Reasoning about his own life seems to
depend upon emotional involvement. Damasio's claim suggests that if
we turn over important policy decisions to computer programs, then
our lives will be a mess because the emotional component was absent in
decision making.

*That would be a threat to Descartes.*

Yes, if what Damasio says is true, it suggests that reason *isn't* separate
from emotion, that reason has everything to do with the capacity for
feeling. That again would shoot down the idea that more logical manipulation would be sufficient to maximize self-interest in a situation. That
is one part of the problem. The other part has to do with understanding. Computers don't understand anything.

*How so?*

They don't have bodies. They cannot experience things. Most of our
abstract concepts are extensions of bodily based concepts that have to do
with motion and space, and objects we manipulate, and states of our bodies, and so on. They then get projected by metaphor onto abstract concepts. We understand through the body. Computers don't have bodies.

This does not mean that important aspects of reason cannot be
modeled on a computer, and indeed I work with people who are engaged
in modeling small aspects of mind. Each small aspect requires a monu-

mental task of analysis and representation, which is not likely to be incorporated into computer technology in anything like the foreseeable future. Perhaps it's so complicated it never will be. But beyond that, there is now no reason whatever to think that the kinds of computations that are done in artificial intelligence programs are "intelligent" in the way that human beings are. All they can do is follow algorithms. Now that does not mean that there is no utility in them — in fact, they can be very useful. But it is important to understand that they are not intelligent in the way human beings are and that they don't understand anything at all.

*But humans can be reduced to doing the kinds of things that can be done "algorithmically," and surely that is what a lot of labor consists of, especially in modern times.*

Yes, that is one of the sad things about industrialization — it tries to turn people into machines. Do computers do that, or do they liberate people from machine-like work? To a significant extent, the computer can turn you into even more of a machine. One of the things that disturbs me in working on computers is certain forms of repetition that make me machine-like, and that's what I simply loathe in interacting with a computer. It could be that future user-friendly computers will eliminate that. I hope so.

But there is another important issue we haven't discussed yet; that is, human limitations. You asked whether it was true that there is more information available to us than before. Well, we cannot possibly process all the information that we could understand. There is no way for a human being to do it. I've had to get off many many e-mail lists simply because, when I get a thousand messages a week, there is no way I can read them. One of the good things about the computer is that it enables people to write more; one of the bad things about the computer is that it enables people to write more; that is, more than you can read.

In many disciplines, largely because of computer technology, more work is produced than anybody can take in. So academic fields are becoming fragmented more and more. Certainly more is being done, but no one can grasp all that is being done or have an overall view of a discipline as was possible twenty years ago. As a result, the new "information" out there is not really knowable. It's not information *for you* — or for another human being. There's only so much one can comprehend.

*So you must find the "information superhighway" metaphor misleading...*

Very misleading. Sure, some things about it seem to make sense. A huge array of things may become potentially available to you directly — lectures, texts, movies, whatever. That's fine, except that every time you take advantage of it, there's something else you can't do. If you think of information as relative to a person, there's only a certain number of waking hours in a lifetime — and you don't want to spend all of them at a computer. Add to that limit the limit on what you can understand and the training it takes to be able to achieve understanding, and there is a strict limit on how much information is available to each person. Already what is available has passed the limit that any person can possibly use. The amount *for you* cannot grow any further. So however many more different *sorts* of things may become available to you, it is not *more*.

*Your account is in terms of bits of information — there is nothing in it of affect or intensity of experience. It seems flattened out.*

You're right. Actually, most of the so-called interactive stuff is pretty uninteractive! It has to do with some fixed menu, not with being able to probe as you would a person or to judge or be moved as you would in a live interaction. There have to be canned answers and canned possibilities. The idea of interactive video is rather minimal now and not likely to be very rich or interesting for a very long time.

*In an ample life, then, how much weight would one attach to technologies such as the computer and video?*

One of the sad things is that the increase in computer technology does not get you out into the world more, into nature, into the community, dancing, singing, and so on. In fact, as the technology expands, there is more expectation that you will spend more of your life at a screen. That is not, for my money, the way one should live one's life. The more that the use of computers is demanded of us, the more we shall be taken away from truly deep human experiences. That does not mean you should never be at a computer screen. Nor does it mean that if you spend time at a computer, you will never have any deep human experiences. It just means that current developments tend to put pressure on people to live less humane lives.

*Less humane, because, for example, at an automatic teller one has to con-form in a mechanical way to the pacing and protocols of a machine?*

Right, you have to conform, and even if you could *say* to the automatic teller, "Machine, give me money," you'd still have to say a form of words, the magic words that will get you the money, and you'd still not be interacting with another human in any sense. Similarly, if you have a computer program that enables you to sing with a recorded orchestra, that is very different from singing with live musicians whom you can groove with — who adjust to you and you to them, and with whom you have a human relationship. What happens is that you get more and more inhuman relationships. That doesn't mean that people using good judgment can't know when to stop.

*But increasingly we do not have the option of unplugging, of saying "Enough!" More and more we are forced into robotic lockstep with some program, such as the interruption by a telemarketing program as it walks down your street at suppertime, on its way through every phone number in your zip code, or having to listen yet again to one's friends' tedious answering-machine messages.*

Yes, I agree, that can be really boring, and although these machines are said to save labor, it's not clear that they do. True, it's easier to cut and paste on a word processor than with paper and scissors. Wonderful, but if you think of all the new demands that come in, it's not clear that you really work less or that your life is more humane.

*Any more than domestic gadgets did to ease the total load of housework for women!*

Well, my mother washed clothes by hand, and that wasn't much fun. But the advent of the washing machine allowed for more demand to be made on both men and women who run a household and take child rearing seriously.

*Given what you have said about the powers and limits of human bodies and the new machines, I take it you find chilling recent speculation about "artificial life."*

This talk about virtual reality and artificial life is at once interesting and silly and weird. Let's start with the positive parts. I could imagine some interesting and fun things to do with virtual reality, and some important ones — for example, ways of guiding surgical operations via virtual reality — so I don't want to put it down. On the other hand, the idea of virtual interactions replacing interactions with real humans or things made of wood, of paper, of natural materials, plants, flowers, animals — *that* I do find chilling. The more you interact not with something natural and alive, but with something electronic, it takes the sense of the earth away from you, takes your embodiment away from you, robs you of more and more of embodied experiences. That is a deep impoverishment of the human soul.

"Artificial life" is a different kind of issue. There is interesting work going on in complexity theory and in the study of what's being called "artificial life." But again, it's being done under certain metaphors, which, like the conduit metaphor, are not always understood as metaphors.

Take the idea, common in the study of artificial life, that life is just the organization of matter, and that the organization can be separated from the thing that's organized. Therefore, if you can represent the organization in the machine, then life would be in the machine. A weird idea. That form of reasoning is metaphorical reasoning, extremely strange metaphorical reasoning, yet a form that seems natural given our metaphorical conceptual system.

There is a very general metaphor called the "properties-as-possessions" metaphor. In expressions like "I have a headache" and "My headache went away," you understand your headache as a possessible object, something that you have, that you can lose. The same headache can even return to you. This metaphor suggests that a headache can exist independently of you — which is a very bizarre idea, a metaphorical entailment, a way of understanding aspects of ourselves as if they were objects.

Similarly, there are aspects of ourselves that are organized, but once you see the organization as a possessible object separable from the organism — which it isn't — then you can think of this property existing independently. Now, thinking that way can be useful — architects think that way. If you isolate the structure of a house, you can draw architectural plans; you can then design buildings more easily. That does not mean, however, that what you have on the plans is the actual structure of a house. The architectural plan is a separate entity, which bears a very indirect relationship to the structure of the house. As soon as you think of the structure of a house as *being* the architectural plan, that is when metaphorical entailment takes over. That is where the mistake is.

The same mistake applies in the understanding of artificial life. If the organization is what gives a thing life, then the life is seen as in the organization. Purely a metaphorical idea. And if organization can be modeled in the computer, and life is in the organization, then the metaphorical logic says that life is in the computer. This is a metaphorical inference made by some people who study artificial life.

*So it's a wild-goose chase, an* ignis fatuus.

Well, it doesn't mean that people shouldn't be studying the organization of living systems. It doesn't mean that complexity theory is not important for the task. But it is not the case that life is in the computer or that the computer is alive.

Take the idea of a computer virus. Is a computer virus alive? Here's a possible argument: The computer virus can reproduce. If it can reproduce, doesn't it have to be alive? But what is called "reproduction" for a computer virus is not, to say the least, the same as reproduction for a human being or even a real virus. It is a metaphor linking reproduction to the idea of copying some structure.

In the study of "genetic algorithms," for example, which models the way certain entities "reproduce," two sequences of bits, say 1100 and 1011, might recombine by taking the first two digits of the first sequence and the last two digits of the second sequence to produce a new sequence 1111. That is called "sexual reproduction" in a genetic algorithm! A metaphor that isn't anywhere near to the actual thing! Researchers in this area go on to speak of "development" and "evolution" in a similarly metaphorical way. It doesn't mean the work is uninteresting or not worthwhile as a technical enterprise. It just isn't sexual reproduction, any more than artificial life is life in the computer.

The justification for the metaphor, of course, comes from a metaphorical understanding of how DNA works. Suppose you've understood both genetic algorithms and DNA in terms of metaphors from the same source domain. That doesn't mean they're the same thing; it means there are two metaphors. And it doesn't mean that there is a new category of life — "artificial life." However, this mode of thought does seem natural, and for good reason.

First, people think metaphorically but they don't know that they do. Secondly, when they think metaphorically, they form new categories, which they believe are real and based on similarity. But metaphors are *not* based on similarity; they are just cross-domain mappings. Once the metaphor is used, it creates a metaphorical similarity. The metaphorical similarity is taken to be a real similarity that defines a new category: a "life" that includes both genetic algorithms and human beings.

What has happened both in the study of artificial intelligence and of artificial life is that researchers are reasoning metaphorically and don't know that they aren't reasoning literally. They think they are finding true existing similarities, when all they are finding is source domains that are the same for different metaphors.

This kind of mistake happens because, first, people use metaphor unconsciously, and, secondly, because you *must* use metaphor to understand most of what what happens in a computer. What happens in a computer can only be comprehended by metaphors; in fact, metaphors are used to design computers. It's that confusion that leads people to talk about artificial intelligence and artificial life.

*What is at stake in this whole discussion of metaphor and the new technologies of information and communication? What, if you like, are the politics in these metaphors?*

There is a great deal at stake both in terms of politics and economics. To begin with economics: the effects on our lives are likely to be enormous. It won't be long before everybody has perhaps half a dozen wires coming into the house, wires they pay for, not just cable TV. The Internet, for example, is not going to be free for very long. There is a very large economic incentive to make people more and more dependent on this technology. Part of the propaganda behind it is that you will have more information at your fingertips. Well, it will be different information, not more information.

*An argument that you have demolished.*

Yes, in the sense that all this information could not possibly be *more* information *for you.* If you have 500 TV channels, how many programs can you watch, even if you wanted to? Then there is the question of who is going to control it. Sometimes that's fine — you and I can put things on the Internet. But advertisers and politicians will, as time goes on, learn to control what is on the Internet in ways they cannot do now.

As you know, I had a remarkable experience putting my paper "Metaphor and War" on the Internet. That was one of the most widely distributed papers ever on the Internet, and it was because, when the Gulf War was about start, there were many people around the world who found that paper useful and they kept forwarding it to recipients on more and more bulletin boards across the Internet. For me, that was a marvelous thing; the paper was read by millions of people.

I suspect that the Internet is now too big for something like that to ever happen again. People are already too jaded. Eventually, much of what will end up on the Internet will be corporate stuff, advertising, entertainment, material from government agencies, and so on. The possibilities for exercising social control are quite remarkable. Take the way Ross Perot tried to set up these community forums around the country, as if they were real community forums. Fifty million people all with access to Perot — that's ridiculous! Perot is there for an hour; how many can ask him a single question, let alone follow up? Twenty? Well, twenty people have "access" to Perot, not fifty million, and he still controls the format. Politicians will want to make this look like a serious form of inquiry. It isn't.

*References*
Damasio, Antonio R. 1994. *Descartes' Error: Emotion, Reason, and the Human Brain.* New York: G. P. Putnam's Sons.
Emmeche, Claus. 1994. *The Garden in the Machine: The Emerging Science of Artificial Life.* Princeton: Princeton University Press.
Lakoff, George. 1992. "Metaphor and War." In *Confrontation in the Gulf.* Edited by Harry Kreisler. Berkeley: Institute for International Studies.
Reddy, Michael. 1993. "The Conduit Metaphor." In *Metaphor and Thought.* 2d ed. Edited by Andrew Ortony. Cambridge: Cambridge University Press.

*Ellen Ullman*

# OUT OF TIME:
# Reflections on the Programming Life

## I.

People imagine that programming is logical, a process like fixing a clock. Nothing could be further from the truth. Programming is more like an illness, a fever, an obsession. It's like those dreams in which you have an exam but you remember you haven't attended the course. It's like riding a train and never being able to get off.

The problem with programming is not that the computer isn't logical — the computer is terribly logical, relentlessly literal-minded. Computers are supposed to be like brains, but in fact they are idiots because they take everything you say completely at face value. You can say to a toddler, "Are yew okay tewday?" But it's not possible for a programmer to say anything like that to a computer. There will be a syntax error.

When you program, your mind is full of details, millions of bits of knowledge. This knowledge is in human form, which is to say rather chaotic, coming at you from one perspective then another, then a random thought, then something else important, then the same thing with a what-if attached. For example, try to think of everything you know about something as simple as an invoice. Now try to tell an idiot how to prepare one. That is programming.

A computer program is an algorithm that must be written down in order, in a specific syntax, in a strange language that is only partially readable by regular human beings. To program is to translate between the chaos of human life and the line-by-line world of computer language. It is an act of taking dictation from your own mind.

You must not lose your own attention. As the human-world knowledge tumbles about in your mind, you must keep typing, typing. You must not be interrupted. Any break in your listening causes you to lose a line here or there. Some bit comes then — oh no, it's leaving, please come back. It may not come back. You may lose it. You will create a bug and there's nothing you can do about it.

Every single computer program has at least one bug. If you are a programmer, it is guaranteed that your work has errors. These errors will be discovered over both short and long periods of time, most coming to light after you've moved to a new job. But your name is on the program. The code library software keeps a permanent record card of who did what and when. At the old job, they will say terrible things about you after you've gone. This is normal life for a programmer: problems trailing behind you through time, humiliation in absentia.

People imagine that programmers don't like to talk because they prefer machines to people. This is not completely true. Programmers don't talk because they must not be interrupted.

This inability to be interrupted leads to a life that is strangely asynchronous to the one lived by other human beings. It's better to send e-mail than to call a programmer on the phone. It's better to leave a note on the chair than to expect the programmer to come to a meeting. This is because the programmer must work in mind-time but the phone rings in real time. Similarly, meetings are supposed to take place in real time. It's not just ego that prevents programmers from working in groups — it's the synchrony problem. To synchronize with other people (or their representation in telephones, buzzers, and doorbells) can only mean interrupting the thought-train. Interruptions mean certain bugs. You must not get off the train.

I used to have dreams in which I was overhearing conversations I had to program. Once, I had to program two people making love. In my dream they sweated and tumbled while I sat with a cramped hand writing code. The couple went from gentle caresses to ever-widening passions, and I despaired as I tried desperately to find a way to express the act of love in the C computer language.

No matter what anyone tells you about the allure of computers, I can tell you for a fact that love cannot be programmed.

## II.

I once had a job where I didn't talk to anyone for two years. Here was the arrangement: I was the first engineer hired by a start-up software company. In exchange for large quantities of stock that might be worth something someday, I was supposed to give up my life.

I sat in a large room with two other engineers and three Sun workstations. The fans of the machines whirred, the keys of the keyboards clicked. Occasionally one or the other of us would grunt or mutter. Otherwise, we did not speak. Now and then, I would have a temper outburst in which I pounded the keyboard with my fists, setting off a barrage of beeps. My colleagues might look up but never said anything about this.

Once a week, I had a five-minute meeting with my boss. He was a heavy-set bearded man with glasses who looked like everyone's stereotype of a nerd; as a matter of fact, he looked almost exactly like my previous boss, another heavy-set bearded man with glasses. At this meeting I would routinely tell him I was on schedule. Since being on schedule is a very rare thing in software engineering, my boss would say good, good, see you next week.

I remember watching my boss disappear down the row of cubbyhole partitions. He always wore exactly the same clothes: he had several outfits, each one exactly the same, khaki pants and a checked shirt of the same pattern. So, week to week, the image of his disappearing down the row of partitions remained unchanged. The same khaki pants, the same pattern in the checked shirt. Good, good, see you next week.

Real time was no longer compelling. Days, weeks, months, and years came and went without much physical change in my surroundings. Surely I was aging. My hair must have grown, I must have cut it, grown more gray hairs. Gravity must have been working on my late-thirties body, but I didn't notice. I only paid attention to my back and shoulders because they seized up on me from long sitting. Later, after I left the company, there was a masseuse on staff. That way, even the back and shoulders could be ignored.

What was compelling was the software. I was making something out of nothing, I thought, and I admit the software had more life for me than my brief love affair, my friends, my cat, my house, my neighbor who was stabbed and nearly killed by her husband. I was creating ("creating," that is the word we used) a device-independent interface library. One day, I sat in a room by myself surrounded by computer monitors from various manufacturers. I remember looking at the screens of my companions and saying, "Speak to me."

I completed the interface library in two years and left the company. Five years later, the company's stock went public. For the engineers who'd stayed the original arrangement was made good: in exchange for giving up seven years of their lives, they became very, very wealthy. As for me, I bought a car. A red one.

III.

Frank was thinking he had to get closer to the machine. Somehow, he'd floated up. Up from memory heaps and kernels. Up from file systems. Up through utilities. Up to where he was now: an end-user query tool. Next thing, he could find himself working on general ledgers, invoices — God — *financial reports.* Somehow he had to get closer to the machine.

Frank hated me. Not only was I closer to the machine, I had won the coin toss to get the desk near the window. Frank sat in full view of the hallway and he was further from the machine.

Frank was nearly forty. His wife was pregnant. Outside in the parking lot (which he couldn't see through my window), his new station wagon was heating up in the sun. Soon, he'd have a kid, a wife who had just quit her job, a wagon with a child-carrier, and an end-user query tool. Somehow he had to get closer to the machine.

Here are the reasons Frank wanted to be closer to the machine: The machine means midnight dinners of Diet Coke. It means unwashed clothes and bare feet on the desk. It means anxious rides through mind-time that have nothing to do with the clock. To work on things used only by machines or other programmers — that's the key. Programmers and machines don't care how you live. They don't care when you live. You can stay, come, go, sleep, or not. At the end of the project looms a deadline, the terrible place where you must get off the train. But, in between, for years at a stretch, you are free: free from the obligations of time.

To express the idea of being "closer to the machine," an engineer refers to "low-level code." In regular life, "low" usually signifies something bad. In programming, "low" is good. Low is better.

If the code creates programs that do useful work for regular human beings, it is called "higher." Higher-level programs are called "applications." Applications are things that people use. Although it would seem that usefulness by people would be a good thing, from a programmer's point of view, direct people-use is bad. If regular people, called "users," can understand the task accomplished by your program, you will be paid less and held in lower esteem. In the regular world, the term "higher" may be better, but, in programming, higher is worse. High is bad.

If you want money and prestige, you need to write code that only machines or other programmers understand. Such code is "low." It's best if you write microcode, a string of zeroes and ones that only a processor reads. The next best thing is assembler code, a list of instructions to the processor, but readable if you know what you're doing. If you can't write microcode or assembler, you might get away with writing in the C or C++ language. C and C++ are really sort of high, but they're considered "low." So you still get to be called a "software engineer." In the grand programmer-scheme of things, it's vastly better to be a "software engineer" than a "programmer." The difference is about thirty thousand dollars a year and a potential fortune in stock.

My office-mate Frank was a man vastly unhappy in his work. He looked over my shoulder, everyone's shoulder, trying to get away from the indignity of writing a program used by regular people. This affected his work. His program was not all it should have been, and for this he was punished. His punishment was to have to talk to regular people.

Frank became a sales-support engineer. Ironically, working in sales and having a share in bonuses, he made more money. But he got no more stock options. And in the eyes of other engineers, Frank was as "high" as one could get. When asked, we said, "Frank is now in sales." This was equivalent to saying he was dead.

# IV.

Real techies don't worry about forced eugenics. I learned this from a real techie in the cafeteria of a software company.

The project team is having lunch and discussing how long it would take to wipe out a disease inherited recessively on the X chromosome. First come calculations of inheritance probabilities. Given some sized population, one of the engineers arrives at a wipeout date. Immediately, another suggests that the date could be moved forward by various manipulations of the inheritance patterns. For example, he says, there could be an education campaign.

The six team members then fall over one another with further suggestions. They start with rewards to discourage carriers from breeding. Immediately they move to fines for those who reproduce the disease. Then they go for what they call "more effective" measures: Jail for breeding. Induced abortion. Forced sterilization.

Now they're hot. The calculations are flying. Years and years fall from the final doom-date of the disease.

Finally, they get to the ultimate solution. "It's straightforward," someone says, "just kill every carrier." Everyone responds to this last suggestion with great enthusiasm. One generation and — bang! — the disease is gone.

Quietly I say, "You know, that's what the Nazis did."

They all look at me in disgust. It's the look boys give a girl who has interrupted a burping contest. One says, "This is something my wife would say."

When he says "wife," there is no love, warmth, or goodness in it. In this engineer's mouth, "wife" means wet diapers and dirty dishes. It means someone angry with you for losing track of time and missing dinner. Someone *sentimental*. In his mind (for the moment), "wife" signifies all programming-party-pooping, illogical things in the universe.

Still, I persist. "It started as just an idea for the Nazis, too, you know."

The engineer makes a reply that sounds like a retch. "This is how I know you're not a real techie," he says.

## V.

A descendent of Italian princes directs research projects at a well-known manufacturer of UNIX workstations. I'm thrilled. In my then five years of being a consultant, the director is the first person to compliment me on what I am wearing to the interview.

It takes me a while, but I soon see I must forget all the usual associations with either Italians or princes. There will be no lovely long lunches that end with deftly peeled fruit. There will be no well-cut suits of beautiful fabrics. The next time I am wearing anything interesting, the director (I'll call him Paolo) tells me I look ridiculous.

Paolo's Italianism has been replaced, like a pod from outer space, with some California New Age, Silicon Valley engineering creature. He eats no fat. He spoons tofu-melange stuff out of a Tupperware container. Everything he does comes in response to beeps emitted from his UNIX workstation: he eats, goes to meetings, goes rollerblading in the parking lot, buys and sells stock, calls his wife solely in response to signals he has programmed into his calendar system. (The clock on his wall has only the number twelve on it.) Further, Paolo swears he has not had a cold since the day he decided that he would always wear two sweaters. Any day now, I expect to see him get out of his stock-option Porsche draped in garlic.

I know that Paolo has been replaced because I have met his wife. We are at a team beer-fest in the local programmer hangout on a Friday afternoon. It's full of men in T-shirts and jeans. Paolo's wife and I are the only people wearing makeup. She looks just the way I expect a no-longer-young Italian woman to look — she has children, she has taken time with her appearance, she is trying to talk to people. Across the swill of pitchers and chips glopped with cheesy drippings, she eyes me hopefully: another grown-up woman. At one point, she clucks at Paolo, who is loudly describing the effects of a certain burrito. "The only thing on earth that instantly turns a solid into a gas," he says.

The odder Paolo gets, the more he fits in with the research team. One engineer always eats his dessert first (he does this conscientiously; he wants you, dares you to say something; one simply doesn't). Another comes to work in something that looks suspiciously like his pajamas. To work on this project, he has left his wife and kids back East. He obviously views the absence of his family as a kind of license: he has stopped shaving and (one can't help noticing) he has stop washing. Another research engineer comes to work in shorts in all weather; no one has ever seen his knees covered. Another routinely makes vast changes to his work the day before deadlines; he is completely unmoved by any complaints about this practice. And one team member screens all e-mail through a careful filter, meaning most mail is deposited in a dead-letter file. This last engineer, the only woman permanently on the project, has outdone everyone on oddness: she has an unlisted work phone. To reach her, you must leave a message with her manager. The officially sanctioned asynchrony of the unlisted phone amazes me. In my fifteen years in the software industry, I have never seen anything like it.

These research engineers can be as odd as they like because they are very, very close to the machine. At their level, it is an honor to be odd. Strange behavior is expected, it's respected, a sign that you are intelligent and as close to the machine as you can get. Any decent software engineer can have a private office, come and go at all hours, exist out of normal time. But to be permanently and sincerely eccentric — this is something only a senior research engineer can achieve.

In meetings, they behave like children. They tell each other to shut up. They call each other idiots. They throw balled-up paper. One day, a team member screams at his Korean colleague, "Speak English!" (A moment of silence follows this outburst, at least.) It's like dropping in at the day-care center by mistake.

They even behave like children when their Japanese sponsors come to visit. The research is being funded through a chain of agencies and bodies that culminates in the Japan Board of Trade. The head of the sponsoring department comes with his underlings. They all wear blue suits. They sit at the conference table with their hands folded neatly in front of them. When they speak, it is with the utmost discretion; their voices are so soft, we have to lean forward to hear. Meanwhile, the research team behaves badly, bickers, has the audacity to ask when they'll get paid.

The Japanese don't seem to mind. On the contrary, they appear delighted. They have received exactly what their money was intended to buy. They have purchased bizarre and brilliant Californians who can behave any way they like. The odd behavior reassures them: Ah! These must be real top-rate engineers!

## VI.

We are attending conventions. Here is our itinerary: we will be traveling closer and closer to the machine. Our journey will be like crossing borders formed by mountain ranges. On the other side, people will be very, very different.

We begin "high," at a conference of computer trainers and technical writers. Women are everywhere. There is a great deal of nail polish, deep red, and briefcases of excellent leathers. In the cold, conditioned air of the conference hall drifts a faint, sweet cloud of perfume.

Next we travel to Washington, D.C., to an applications development conference, the Federal Systems Office Expo. It is a model of cultural diversity. Men, women, whites, blacks, Asians — all qualified applicants are welcome. Applications development ("high-level," low-status, and relatively low-paying) is the civil service of computing.

Now we move west and lower. We are in California to attend a meeting of SIGGRAPH, the graphics special interest group of the Association of Computing Machinery (ACM). African Americans have virtually disappeared. Young white men predominate, with many Asians among them. There are still some women: graphics can be seen, after all. Though we have crossed the summit and have begun our descent, we are still not very "low."

On our map, we must now place this warning: "Below here be engineers."

We are about to descend rapidly into valleys of programming, to the low levels close to the machine. We go first to an operating-systems interest group of the ACM. Then, getting ever closer to hardware, we attend a convention of chip designers. Not a female person in clear sight. If you look closely, however, you can see a few young Chinese women sitting alone, quiet, plainly dressed, succeeding at making themselves invisible. For these are gatherings of young men. This is the land of T-shirts and jeans, the country of perpetual graduate-studenthood.

Later, at a Borland developers' conference, company engineers proudly call themselves "barbarians" (although they are not really as "low" as they think they are). In slides projected onto huge screens, they represent themselves in beards and animal skins, holding spears and clubs. Except for the public-relations women (their faint clouds of perfume drifting among the hairy, exposed barbarian legs), there is only one woman (me).

A senior engineer once asked me why I left full-time engineering for consulting. At the time, I had never really addressed the question, and I was surprised by my own answer. I muttered something about being a middle-age woman. "Excuse me," I found myself saying, "but I'm afraid I find the engineering culture very teenage-boy puerile."

This engineer was a brilliant man, good-hearted, and unusually literate for a programmer. I had great respect for him and I really did not mean to offend him. "That's too bad," he answered as if he meant it, "because we obviously lose talent that way."

I felt immense gratitude at this unexpected opening. I opened my mouth to go on, to explore the reasons for the cult of the boy engineer.

But immediately we were interrupted. The company was about to have an interdivisional water-balloon fight. For weeks, the entire organization had been engaged in the design of intricate devices for the delivery of rubberized inflatable containers filled with fluid. Work had all but stopped; all "spare brain cycles" were involved in preparations for war.

The engineer joined the planning with great enthusiasm, and I left the room where we had been having our conversation. The last I saw of him, he was covering a paper napkin with a sketch of a water-balloon catapult.

Here is a suggested letter home from our journey closer to the machine: Software engineering is a meritocracy. Anyone with the talents and abilities can join the club. However, if rollerblading, Frisbee playing, and water-balloon wars are not your idea of fun, you are not likely to stay long.

### VII.

I once designed a graphical user interface with a man who wouldn't speak to me. My boss hired this man without letting anyone else sit in on the interview; my boss lived to regret it.

I was asked to brief my new colleague and, with a third member of the team, we went into a conference room. There, we filled two whiteboards with lines, boxes, circles, and arrows in four marker colors. After about half an hour, I noticed that the new hire had become very agitated.

"Are we going too fast?" I asked him.

"Too much for the first day?" said the third.

"No," said our new man, "I just can't do it like this."

"Do what?" I asked. "Like what?"

His hands were deep in his pockets. He gestured with his elbows. "Like this," he said.

"You mean design?" I asked.

"You mean in a meeting?" asked the third.

No answer from our new colleague. A shrug. Another elbow gesture.

Something terrible was beginning to occur to me. "You mean talking?" I asked.

"Yeah, talking," he said. "I can't do it by talking."

By this time in my career, I had met many strange engineers. But here was the first one who wouldn't talk at all. Besides, this incident took place before the existence of standard user interfaces like Windows and Motif, so we had a lot of design work to do. No talking was certainly going to make things difficult.

"So how *can* you do it?" I asked.

"Mail," he said immediately, "send me e-mail."

So, given no choice, we designed a graphical user interface by e-mail.

Corporations across North America and Europe are still using a system designed by three people who sent e-mail and one who barely spoke.

## VIII.

Pretty graphical interfaces are commonly called "user-friendly." But they are not really your friends. Underlying every user-friendly interface is a terrific human contempt.

The basic idea of a graphical interface is that it does not allow anything alarming to happen. You can pound on the mouse button all you want, and the system will prevent you from doing anything stupid. A monkey can pound on the keyboard, your cat can run across it, your baby can fist it, but the system should not crash.

To build such a crash-proof system, the designer must be able to imagine — and disallow — the dumbest action. He or she cannot simply rely on the user's intelligence: who knows who will be on the other side of the program? Besides, the user's intelligence is not quantifiable; it's not programmable; it cannot protect the system. No, the real task is to forget about the intelligent person on the other side and think of every single stupid thing anyone might possibly do.

In the designer's mind, gradually, over months and years, there is created a vision of the user as imbecile. The imbecile vision is mandatory. No good, crash-proof system can be built except it be done for an idiot.

The designer's contempt for your intelligence is mostly hidden deep in the code. But, now and then, the disdain surfaces. Here's a small example: You're trying to do something simple like copy files onto a diskette on your Mac. The program proceeds for a while then encounters an error. Your disk is defective, says a message, and, below the message, is a single button. You absolutely must click this button. If you don't click it, the program hangs there indefinitely. So, your disk is defective, your files may be bollixed up, and the designer leaves you only one possible reply: You must say, "OK."

The prettier the user interface, and the fewer odd replies the system allows you to make, the dumber you once appeared in the mind of the designer.

## IX.

The computer is about to enter our lives like blood in the capillaries. Soon, everywhere we look, we will see pretty, idiot-proof interfaces designed to make us say, "OK."

A vast delivery system for retail computing is about to come into being, and the system goes by the name "interactivity." Telephones, televisions, sales kiosks will all be wired for interactive, on-demand services. The very word — interactivity — implies something good and wonderful. Surely a response, a reply, an answer is a positive thing. Surely it signifies an advance over something else, something bad, something that doesn't respond, reply or answer. There is only one problem: what we will be interacting with is a machine.

Interactive services are supposed to be delivered "on demand." What an aura of power — demand! See a movie, order seats to a basketball game, make hotel reservations, send a card to mother — all services waiting for us on our television or computer whenever we want them. Midnight, dawn, or day. Sleep or order a pizza: it no longer matters exactly what we do when. We don't need to involve anyone else in the satisfactions of our needs. We don't even have to talk. We get our services when we want them, free from the obligations of regularly scheduled time. We can all live closer to the machine.

"Interactivity" is misnamed. It should be called "asynchrony": the engineering culture coming to everyday life.

In the workplace, home office, sales floor, service kiosk, home — we will be "talking" to programs that are beginning to look surprisingly alike: all full of animated little pictures we are supposed to pick, like push-buttons on a toddler's toy. The toy is supposed to please us. Somehow, it is supposed to replace the satisfactions of transacting meaning with a mature human being, in the confusion of a natural language, together, in a room, at a touching distance.

As the computer's pretty, helpfully waiting face (and contemptuous underlying code) penetrates deeply into daily life, the cult of the boy engineer comes with it. The engineer's assumptions and presumptions are in the code. That's the purpose of the program, after all: to sum up the intelligence and intentions of all the engineers who worked on the system over time — tens and hundreds of people who have learned an odd and highly specific way of doing things. The system contains them. It reproduces and reenacts life as engineers know it: alone, out-of-time, disdainful of anyone far from the machine.

Engineers seem to prefer the asynchronous life, or at least be used to it. But what about the rest of us? A taste of the out-of-time existence is about to become possible for everyone with a television. Soon, we may all be living the programming life. Should we?

*John Simmons*

# SADE AND CYBERSPACE

### 1.

Max Horkheimer and Theodor Adorno (1947) suggest in Excursus II of *The Dialectic of Enlightenment* that the Marquis de Sade's *Juliette* is a central text for examining the character of capitalism and emergence of fascism, in terms of reason and the manipulation of rational thought. They are not alone: *L'Histoire de Juliette*, remarks Geoffrey Gorer (1963, 126), "is one of the most thorough analyses, as it is by fifty years the first, of a society ruled by money." The novel's philosophical breadth and extraordinary resistance to recuperation render it an enduring instrument for social critique even apart from the great beauty of its language which makes it a masterpiece of French literature. Indeed, the fact that it is a novel at all needs to be understood carefully: for Sade, as Alice Laborde (1974, 13) rightly asserts, "the novel has a mission: It is not a question of a type of *divertissement* which permits escape from the realities of life but a means of subtle revelation, of unlimited suggestive powers."

### 2.

That the Marquis de Sade bears importance for cyberspace should not be surprising. A thoroughgoing materialist, Sade admired La Mettrie's version of "man a machine," which stressed the interdependence of mind and body, and he would be amused, if not outraged, by the persistence of the Cartesian precepts that inform cyberspace and render plausible such dystopic technological appropriations as "artificial intelligence." Sade offers a broader horizon of thought and action by which to comprehend cyberspace, especially when it is brought into domains where it does not belong. For example, we find the notion of feedback in Sade as that which arises from simultaneous and contradictory desires to solicit and control stimulation. "Be still," Norceuil demands of the unfortunate Laïs,

> Stop that squirming . . . do you not see what the imagination is like for a man like me? Anything can disturb and disrupt it; as soon as we fail to serve it, it breaks down, and the most divine charms are worthless unless presented to us submissively and with obedience. (Sade 1797, VI) (My translation, here and throughout.)

A further reason for Sade's relevance is that in his work he is committed — as are we, presumably — to the possibility that human reason may affect conduct and destiny. He therefore provides a critique of all forms of control: "The triumph of philosophy," he writes in the famous first sentence of *Les Infortunes de la vertu,*

> would be to cast light into darkness along the ways used by Providence to bring about the ends it sets for man, and to trace subsequently some plan of conduct that might make known to this unfortunate biped, perpetually buffeted by the caprices of the Being that is said to rule over him so tyrannically, the manner in which he must interpret the decrees of this Providence, the route to which he must tend, in order to avoid the bizarre caprices of this fatality to which are given 20 different names without yet coming to define it. (Sade 1791)

If there is angst in cyberspace, and a gulf between dream and the reality, it lies here, in the promise apposite the failure to provide.

3.
In what follows I seek to show how, from a Sadean perspective, computers are, when used for purposes other than transmission of low-con-

text messages (Hall 1974), frequently *noncorporeal excitement generators.* (High- or low-context messages: to what extent meaning is dependent upon or embedded in context rather than the message itself. A train schedule is low context, for example; all signficant art is high context.) This helps to account for the extraordinary triviality that is so impressive about cyberspace: The fact is that, for all that computers generate the rhetoric of change and even "revolution," their cultural impact to date, however prodigious, apart from often dolorous changes in the workplace, is largely in the realm of games (and so they are also *aggression absorption facilitators*) and of meeting false needs. This is in great part because cyberspace tends toward religious space, which accounts for its hypnotic fascination and aptitude in occupying the imagination.

Insight provided by a reading of Sade does not point toward Luddism but rather toward recognition that an alternative, Epicurean vision can perhaps be conceived in the context of cyberspace and virtual, or what is called here *prosthetic,* reality.

4.

Today cyberspace is invading the psyche outside the workplace, much as the computer made efforts to occupy the shop floor over the past generation. Its initial results are often unsuccessful and clumsy, and it is tempting to dismiss entirely the odd and amusing mixture of hyperbole and philosophy that emanates from Silicon Valley. But this would be a mistake, not only because power lies there and not with academic philosophers, but also because, as Horkheimer and Adorno suggested in 1947, come the Enlightenment and capitalism in its wake, "Being is apprehended under the aspect of manufacture and administration." It is then of enormous importance if these latter terms significantly change their *modus operandi.* Cyberspace promises such change, extending from the workplace to marketplace, which both grow increasingly invasive. "The conceptual apparatus determines the senses, even before perception occurs," write Horkheimer and Adorno in . "*Juliette* or Enlightenment and Morality":

> Intuitively, Kant foretold what Hollywood consciously put into practice: in the very process of production, images are precensored according to the norm of the understanding which will later govern their apprehension. (1947, 84)

With the arrival of computer-driven technology following World War II, it was perhaps sufficient to use military metaphors to emphasize the machine's ultimately inhuman rationality and capability to dominate (Levidow and Robins 1989). But today it is important to realize that cyberspace would change the conceptual apparatus itself, and it is critical to understand how and under what rubrics this will be accomplished. And on that account it is especially rewarding to turn — as did Horkheimer and Adorno, seeking to understand totalitarianism, half a century ago — to the Marquis de Sade.

## 5.

More clearly than any previous critic, Annie Le Brun (1986, 1989) has conceptualized what I take to be the underlying organizer in Sadean thought — namely, "les rapports entre la tête et le corps." With the body absent — or guarded apart by moral and religious conviction — arises ideology. "The distinctive quality of ideologies," writes Le Brun, apropos of *Juliette*, "is the production of ideas without bodies, of ideas which are only developed to the detriment of the body" (1989, 20). Sade's attack on religion and the proclaimed rationality of much Enlightenment thinking, both of which in the end seek to control thought and action, is not only direct but given great force by its blending of sexual and philosophical rhetoric.

Le Brun's formulation of this basis of Sade's thinking is fundamental to understanding his deeper significance for the history of Western thought and his continuing importance today. It can be seen in less elegant forms elsewhere; for example, in Roland Barthes's view that Sade's writing sets itself a task: "Sa tâche, dont elle triomphe avec un éclat constant, est de contaminer réciproquement l'érotique et la rhétorique" (1971, 38). But Annie Le Brun has stated Sade's case most clearly and examined some of its broader implications.

## 6.

*Sade: A Preliminary Remark.* To understand Sade in cyberspace it is necessary to know something about Sade, and ignorance and misunderstanding are not merely stumbling blocks: they are cultural impediments born of the very process just outlined (see 4 above). *L'Histoire de Juliette* was a well-known and best-selling book in its day, before its images were subject to perceptual precensorship. At the root of resistance to understanding Sade is a horror of actually reading him; and the extent to which this is the case, even for social critics who ought to know better, is nothing short of astonishing.

The Sadean critique is not self-evident, especially if the author goes unread and unexperienced. But apart from the generation of problematic French critics now passing, which includes Georges Bataille and Maurice Blanchot, it should be noted that adequate material exists in English, some of which dates from before the Second World War. Geoffrey Gorer's 1963 *The Life and Ideas of the Marquis de Sade* is a revised edition of a volume originally published in 1934. The intellectual historian Lester Crocker in *An Age of Crisis* was already explicit, in 1959, concerning Sade, "whose important place in the thought of his age has been shamefully neglected," and he wrote that, "It is not too much to say that the crisis of modern culture is crystallized in Sade" (10, 11). The only truly extensive bibliography of Sade was published in the United States in 1986, by Colette Verger Michael; it remains unavailable in France. Annie Le Brun's magisterial *Sade: A Sudden Abyss* was translated and published here in 1990.

This is not the place to dwell on the mystification that has shaped the image of Sade for an English-speaking audience. It should be noted, however, that while the long-forbidden Sadean corpus was translated and published in the 1960s, Sade's great philosophical and epistolary novel *Aline et Valcour* has never been put in English. And it is arguable that one of the great editorial mistakes of all time was to affix to the Grove Press edition of *The 120 Days of Sodom* a preface by Simone de Beauvoir, thus insuring that the least philosophical or accessible, and the most scatological text — withal of great beauty, some of which is necessarily lost in translation — was the one that many American readers purchased. And finally, it should be recalled to American poststructuralists that Michel Foucault held in high contempt those who were unversed in Sade (Miller 1993).

This is all to say that the substance of avoidance when it comes to Sade takes the form primarily of a psychological resistance; and, indeed, some psychoanalytic insight must be incorporated into any contemporary exegesis. Sadean thought works well with object-relations theory; it is unencumbered by the concept of instinct and so remains refreshingly modern. Like certain other Enlightenment figures, Sade adumbrates Darwin, under whose influence Freud developed the drive theory that, ultimately, long sequestered analytic thinking.

**7.**

*Sadean Space and Cyberspace.* With Sade (unlike other Enlightenment thinkers), it is both plausible and useful to incorporate the concept of space. Sade's imprisonment was crucial to his work, and it is from cells in the Château de Vincennes and the Bastille that his thought is first elaborated. "L'Auteur croit devoir prévenir qu'ayant cédé son manuscrit lorsqu'il sortit de la Bastille," wrote Sade (1793, 23) in the prefatory "Essentiel à Lire" to *Aline et Valcour*, "il a été, par ce moyen, hors d'état de le retoucher." In addition, Sade's passion was theater and his vision, basically theatrical. His great *La Philosophie dans le Boudoir* is in play form with the long discourse, "Français, encore un effort si vous voulez être Républicains," as a speech written by the principal, Dolmancé. The episodes in *The 120 Days of Sodom* are recounted in a theatrical space that Roland Barthes has attempted to reproduce in his *Sade, Loyola, Fourier;* as Barthes points out, the Château de Silling of *The 120 Days* is "hermetiquement isolé du monde par une suite d'obstacles qui rappellent assez ceux que l'on trouve dans certaines contes de fées."

Space effectively enabled Sade to argue the relationship between thought and action, and he may be seen as the most philosophical of novelists but the least abstract of philosophers — rearguing and challenging while keeping in view the distinction between reflecting upon and doing in an extremely modern way every proposition of the Enlightenment. In *Juliette*, Sade's most extensive work, real and existing locations are used, and usually turned into sites for sexual and philosophical orgies: the convent at Panthemont where Juliette is raised; St. Peter's Basilica, where she cavorts and argues with the Pope; and the boudoir of King Ferdinand of Naples are just a few examples. In *Juliette*, it is not too much to say that *Sadean thought occupies and liberates essentially oppressive space.* Sade liberates it from the fantasies of religion, the oppression of women, the hypocrisy of men, and the weight of tyrants.

**8.**

Considered textually, *Juliette* is a series of sexual and "criminal" episodes with philosophical discussions as long interludes. Critics have often been unclear as to the relation between what appears at first as scenes of effusive pornography and the protracted "digressions" of philosophical argument: as recently as 1972 Vera Lee described "The Sade Machine" as "a visual description of vice [alternating] with the verbal justification of vice. The physical and the abstract succeed each other with see-saw regularity." However, recognition of some closer connection between the "physical and the abstract" dates as far back as the rediscovery of Sade for the twentieth century. "I . . . found that, if the obscenity can be, not overlooked, but taken in one's stride," wrote Geoffrey Gorer, "a view of the world was presented of great originality and curious force" (1963, 11). Indeed, the power of the Sadean text — as well as the resistance to reading it — is due to the fact that sex, crime, and philosophy occupy the same space. Mondor, the God-fearing murderer, sodomizes and finally kills an eleven-year-old boy, rips out his heart and rubs it on his face while he ejaculates. No sooner finished, he *quits the room,* which prompts Juliette to remark that

> such are the effects of libertinage on timid souls: remorse and shame follow straight upon the throes of orgasm, because such people, unable to forge their own principles, always imagine themselves to have done wrong simply because they have not behaved like everybody else. (Sade 1797, III)

Juliette herself would never behave this way. Her great debauch with the pope is a gem of blasphemy. When she meets Braschi, he takes her to

> a remote chamber where indolence and luxury, under the drab banner of religion and modesty, offered to salacity all that might flatter its propensities. All mingled together: next to Theresa in ecstasy one saw Messalina embuggered, and beneath an image of Christ, Leda. . . ." (Sade 1797, IV)

After upbraiding him and cataloging the church's insults to humanity, she allows him to give her a guided tour of the palace before helping him onto a sofa and taking measures that "put [her] in a position to analyze the Holy Father." The ensuing analysis by philosophical and anthropological essay — which includes, I perhaps should add, a defense of a woman's right to control her body — concludes with an orgy in St. Peter's Basilica.

These examples could be multiplied a hundredfold in *Juliette* and elsewhere in Sade. The mode of presentation accounts for the fact that the text cannot be recuperated — and, indeed, it is still despised in many quarters, where it also goes unread. Sade's insistence on the significance of the body not only prefigures Freud, but brings to the fore an aspect of Enlightenment thought (recognition of the body) that earlier philosophers (including Hume and the English skeptics) acknowledged in principle but preferred not to think about. As Lester Crocker (1959, 11) points out, "Sade only draws the ultimate conclusions . . . from the radical philosophies developed earlier in the century." But in doing so, his thought functions *à plein régime*, as Annie Le Brun (1989, 45) puts it: "Et cette tête fonctionne à plein régime, parce qu'il y a un corps derrière, parce que ce corps et cette tête, pour Sade, n'existent jamais l'un sans l'autre."

And this, I maintain, provides a fundamental context in which to examine cyberspace.

## 9.

*Cyberspace.* The deemphasis on the body characteristic of cyberspace represents necessity as virtue. In *The Human Use of Human Beings,* Norbert Wiener (1954) does not so much discuss the implications of the body in feedback control, as to create a context in which they do not matter. The whole of chapter 5, "Organization as the Message," is an extraordinary artifice. "The present chapter will contain an element of phantasy," Wiener writes, adding that, "Phantasy has always been at the service of philosophy." He reprises the concept of homeostasis, writing that "We are but whirlpools in a river of ever-flowing water. We are not stuff that abides, but patterns that perpetuate themselves." His aim is to conceptualize the human being as a message that hypothetically can be transmitted. And then he calls upon a story by Rudyard Kipling, "The Night Mail," which emphasizes the powerful importance of the invention of the airplane and development of aviation. "[Kipling] does not seem to realize that where a man's word goes, and where his power of perception goes, to that point his control and in a sense his physical existence is extended." Finally, Wiener advises us to "reconsider Kipling's test of the importance of traffic in the modern world from the point of view of a traffic which is overwhelmingly not so much the transmission of human bodies as the transmission of human information."

The disembodied body of cyberspace, the body reduced to information, was a necessary artifice. Cybernetics insisted on treating in the same theory the human being and the machine, in which the body was a significant encumbrance (Heims 1980). Wiener himself was unprepared for the fantasies of omnipotence that this engendered among militarists and industrialists, and he consciously refused to play a part in their adventures. Among other things to which he turned was the application of cybernetics to the design of replacement parts for the human body — prosthetics.

**10.**

Of course, the disembodied body does not originate with cybernetics, which only extends the subterfuges of Western thought in new directions. It is a false construct that forms one basis of what David Noble (1984, 351) calls "the irrational and infantile ideology of technological progress." Although there is no room for the male or female body in cyberspace, *it abides as an abstraction and is present in all the exploitative uses of computers.* Phantasy indeed comes to the service of philosophy — an instrumental philosophy that is isolated from the body — and occupies the imagination as an alien force. Cyberspace in that sense may become exciting — though not with the insistent interplay between the head and the body (the physical and emotional and intellectual structures) — demanded in Sadean space. Computer games thereby operate as *excitement generators* that take advantage of the *absence* of physical consequences in cyberspace for their impact.

Indeed, the widespread popularity of computer-based games represents a fundamental and rampant confusion as to what constitutes pleasure, and descriptively resembles compulsion neurosis as described by Otto Fenichel (1945) half a century ago. Where Sadean space — in which the text simultaneously generates emotions and philosophy, manipulating both organs and rational thought — depends on the persistence of infantile omnipotence to overcome inhibition due to proscriptions upon the physical structure, in cyberspace such omnipotence is used defensively to erect forbidden zones, embark on wild-goose chases, and indulge in fantasies of swords and sorcerers. The uses to which bodies are put in Sade accounts for the sublime beauty of his texts; their absence (or abstract presence) in cyberspace explains how computers are "exciting" while at the same time emotionally void.

**11.**

"The architectonic structure of the Kantian system, like the gymnastic pyramids of Sade's orgies," write Horkheimer and Adorno, reveals an "organization of life as a whole which is deprived of any substantial goal." This consequence — historically, of the degradation of religion — in cyberspace does not go unnoticed. And, indeed, the reinsertion of *religiosity* into cyberspace is perhaps its most outstanding feature.

This concept — cyberspace as essentially religious — must be examined critically, for it is too likely to be an artifact of Sadean space, which is naturally blasphemous. But there is no need to dwell on religiosity: cyberspace is *acorporeal spiritual space*. It is space where one may work or play in solitude, alienated not simply from others but from one's own physical and emotional self. Feedback is not the vital aspect of cyberspace if you consider that alt.sex.bondage and similar on-line forums are essentially confessional ("He rolled around and begged, 'Please! Please, no more!'"), and it is no wonder that consensuality of behavior is a frequent topic of debate. ("I have examined my life as a slave, and found that for me it is a wonderful and empowering experience") (Kadrey 1994). When the body does appear, it is in the context of taboo and transgression, under the direction of, say, Bob Guccione, with "Penthouse's Virtual Photo Shoot" ("Hi, I'm Dominique . . . Let's get interactive . . . Play w/butt") (Berry 1994). Whether the output of Necro Enema Amalgamated represents an assault on cyberspace or an accommodation does not much matter; it *(Blam!)* is what Hegel called an abstract negation, a magazine of a new order but with the oldest sphincter operation known to man. All such phenomena represent cyberspace as an environment top-heavy with repression and trivialized by the persistence of shame.

**12.**

Given the context provided by Sadean space, if we are willing to engage the crude vulgarism to which cyberspace currently tends, its meaningless and oppressive character becomes evident. According to Jon Carroll (1994), the first "smash hit" CD-ROM is "Myst," an interactive game that has been praised in *Rolling Stone* and, because it is not overtly violent, holds great appeal for parents. But "Myst," according to Carroll, is much more; it may be "the first interactive artifact to suggest that a new art form may well be plausible, a kind of puzzle box inside a novel inside a painting, only with music." "Myst" is a game, albeit "a phenomenon like no other"; it has won several awards and is "beautiful, complicated, emotional, dark, intelligent, absorbing." The game has no clear premise, but "In a vague sort of way, the player understands the task: find out what happened" and its themes are "sound, water, gears, energy." As in other computer games, players pick things up off the virtual ground but, mostly, "they just use their minds."

Most of this is may be ignored as hyperbole, but the insistence on beauty and art is significant. "Myst" is not an excitement generator so much as something that adumbrates a new art form, or wants to. As Carroll states it, "mostly it does not dazzle; it is like careful writing, always furthering the plot or mood." So we ought not be surprised to learn: "Like others before them (Dante, Milton, Blake), the Millers encountered their dark sides even while searching for the light."

Dante? Milton? Blake? Their dark sides? The creators of "Myst" are sons of a peripatetic "nondenominational preacher." As youths they moved often with their father as he found flocks "wherever the Lord needed him and a congregation was available to pay his modest salary." He was a liberal preacher who encouraged debate, according to Carroll, and the Miller boys were free to ask questions and to say what they thought: "free to question" and "free to roam." Rand, the eldest brother, was trained as a computer programmer — obligation of the name! — and introduced his brother to cybernetic games, and "a partnership was formed." Subsequently, one Sunday at church, the Millers met the individual — a carpenter! — who became their business manager. And thus was born "Myst."

Carroll asks the Miller brothers about their religious convictions. He wonders how creating "a whole world" has changed their idea of God. Rand replies that this experience "just makes us realize how great God is." His brother, Robyn, presented with humble pie, demurs somewhat, perhaps because the meek do not always inherit the earth, saying rather that "sometimes late at night, after I had done something really cool, I would look down on my creation and I would say, 'It is good.'"

Carroll himself best articulates the ideology immanent in "Myst": the computer, he writes, is "the engine of democratization" and the Millers are "new players in a new game."

It's not just Silicon Valley anymore. . . . Anyone with guts and talent can be a player; lines of code don't ask about religion, political opinions, taste in clothing or music. We are used to the idea that rebels can find the cracks in the new systems; we are not used to the idea that rebellion doesn't matter anymore. It's pure imagination, unfettered by trend or anti-trend; it can happen anywhere the hardware lives.

This sermon evokes the basic appeal of Christianity (the merest fool can belong), the insignificance of rebellion (in light of spirituality), and most particularly the pure (unconflicted) imagination that imagines what it is told to imagine and no more. In Sadean space, by contrast, lines of code (views of religion and politics) are extremely important: the engines of democracy assume quite other dimensions and forms; and the hardware in question has kinaesthetic and glandular properties currently unavailable to computers. The imagination depends upon and operates on account of them. Citing La Mettrie ("How abundantly happy . . . are those whose lively and wanton imagination keeps their senses ever primed to the foretaste of pleasure"), Belmor discusses this (authentically) unfettered imagination. "Truly, Juliette," he says,

> I do not know if reality measures up to the images we fashion from it, or if the pleasures one does not possess are not worth a hundredfold what is ours: I behold before my eyes your ass, Juliette; I find it beautiful, but my imagination, still more brilliant than nature, and more skillful, I daresay, creates others more beautiful still. And is not the pleasure I gain from this illusion preferable to that which reality brings me? . . . it seems to me that I might do things with this ass of my imagination that the gods themselves would never invent.

These observations demonstrate how the imagination is not only dependent on the physical structure, but that it at the same time shapes the perception of the body itself. In this context — Sadean space — pleasure assumes a new positive value in a contemplative register, giving new meaning to the interrelationship between liberty and beauty. No rapture, and certainly no hyperbole, is called for.

**13.**

*Religiosity and Omnipotence.* Not surprisingly, there is an effort to create an omnipotent, purely spiritual being in cyberspace, and I mean, of course, the pursuit of artificial intelligence. Withal this is a complex phenomenon with a considerable history, I want only to point out that Sade would be more than willing to propose his own Turing-like test; and it would not consist, exactly, in answering questions posed by someone or something in another room. In fact, there is a delightful and amusing divination game, which Juliette plays in Florence with the Duke of Pienza. It does not involve cold reasoning but rather music — the abstract emblem of sensuality. In Sadean space the Turing test becomes a Cartesian curiosity in which "intelligence" and "intellectual feat" are revealed as absurd pretensions. (Note that Sadean space does not insult other scientific thought experiments: Einstein's special theory of relativity invokes bodies in space; quantum physics undercuts the notion of cause and effect.) To be sure, given Turing's difficult and ultimately tragic life story (Hodges 1983) — he was the subject of an invasive state and what he did with his body was at issue — it is not surprising that he developed a test in which the body (or computer) solicited questions while cloistered in another room. What is surprising is that this proposed hypostatic union has been taken seriously.

**14.**

*Final Observations.* With cyberspace — and virtual reality — come new combinations of media that are psychologically invasive. Cyberspace no longer merely offers to bear information that must be fitted into a personally created context; it provides context, however primitive. Cyberspace infantilizes even while it enlightens, appealing to the senses with speed and power and exploiting receptors of human emotion with vivid new colors and kinaesthetic awareness. In doing so, it provokes confusion over the means by which we — in the limited sense in which we are a *we* — conceptualize experience, examine self and others, and generate and perceive beauty.

Hyperbole and cultural pretension — a combination highly prized in the United States — should not diminish our awareness of the potential significance of cyberspace. I would think that Sadean space, in fact, underscores the effectiveness of using cyberspace to transmit and manipulate low-context messages of all kinds, and locates it historically on a continuum with the transformations at the end of the last century — electric light, color printing, the recording of sounds and images. Indeed, just as the last century's proliferation of new forms of leisure and "a joyous vision of comfort" (Susman 1984) brought into relief the extent of mental oppression, today cyberspace underscores anew the limitations of this continuing cornucopia.

Nothing unusual, then, to finally wonder if a plausible and alternative response to cyberspace would not be Epicurean. Faced with the outpouring of insistent, demanding trivia parading as something else, it may be past time to cultivate an ataraxic response as well as to make demands on the world about to be wired: to abandon the search for artificial intelligence; to recognize the severe limitations of such electronic fugues as virtual — prosthetic — reality; to cease to invest pretensions to omnipotence in the computer — in short, an attack on the religiosity of cyberspace. It is not a bad idea, in other words, to turn back to Lucretius, like the Marquis de Sade — like Einstein, if you prefer — and recognize the consequences of reverence, whether for the Pantheon or that fated "engine of democracy":

> *Poor humanity, to saddle the gods with such responsibilities and throw in a vindictive temper! What griefs they hatched then for themselves, what festering sores for us, what tears for our posterity!*

### References
Barthes, Roland. 1971. *Sade, Fourier, Loyola.* Paris: Seuil.
Berry, Colin. 1994. "Dear Penthouse." *Wired* (May).
Carroll, Jon. 1994. "Guerrillas in the Myst." *Wired* (August).
Crocker, Lester. 1959. *An Age of Crisis.* Baltimore: Johns Hopkins University Press.
Fenichel, Otto. 1945. *The Psychoanalytic Theory of Neurosis.* New York: Norton.
Gorer, Geoffrey. 1963. *The Life and Ideas of the Marquis de Sade.* New York: Norton.
Hall, Edward. 1974. *Beyond Culture.* New York: Doubleday.
Heims, Steve J. 1980. *John von Neumann and Norbert Wiener.* Cambridge: MIT.
Hodges, Andrew. 1983. *Alan Turing: The Enigma.* New York: Simon & Schuster.
Horkheimer, Max, and Adorno, Theodor. 1947. "*Juliette* or Enlightenment and Morality." In *Dialectic of Enlightenment.* Translated by John Cumming. New York: Seabury Press 1972.
Kadrey, Richard. 1994. "alt.sex.bondage." *Wired* (June).
Laborde, Alice. 1974. *Sade Romancier.* Neuchâtel: Editions de la Baconnière.

Le Brun, Annie. 1989. *Sade, aller et détours*. Paris: Plon.

―――. 1986. *Soudain, un bloc d'abîme, Sade*. Paris: Pauvert. [*Sade: A Sudden Abyss*. Translated by Camille Naish. San Francisco: City Lights 1990.]

Lee, Vera. 1972. "The Sade Machine." *Studies in Voltaire and the Eighteenth Century*, 98.

Levidow, Les, and Kevin, Robins, eds. 1989. *Cyborg Worlds: The Military Information Society*. London: Free Association.

Lucretius. *On the Nature of the Universe*. Translated by R. E. Latham. New York: Penguin 1951.

Michael, Colette Verger. 1986. *The Marquis de Sade: The Man, His Works, and His Critics: An Annotated Bibliography*. New York: Garland.

Miller, James. 1993. *The Passion of Michael Foucault*. New York: Simon & Schuster.

Noble, David F. 1984. *Forces of Production*. New York: Knopf.

Sade, D.-A.-F. 1791. *Les Infortunes de la vertu*. Paris: Editions Fourcade 1930.

―――. 1793. *Aline et Valcour, ou le Roman philosophique*. Paris: Pauvert 1966–67.

―――. 1797. *Histoire de Juliette*. Paris: Pauvert 1966–67.

Susman, Warren. 1984. *Culture as History*. New York: Pantheon.

Wiener, Norbert. 1954. *The Human Use of Human Beings: Cybernetics and Society*. New York: Doubleday.

DEGRADING

*Doug Henwood*

# INFO FETISHISM

B ack in the summer of 1987, when the Eighties were at their Roaring-
est, an interview with George Gilder ran on the now-defunct
Financial News Network. Gilder, looking like he'd just beamed aboard
from Melville's *Fidèle* (the flagship of *The Confidence-Man*), argued that
the trade deficit was nothing to worry about. Trade figures count only
things, said the goofy poet laureate of entrepreneurship, but what really makes the world move today is information: today, capital bounces
around on satellites and dances up and down fiber-optic cables. Oddly,
Gilder treated the terms "information" and "capital" almost as if they
were synonyms.

Two years later, Gilder published *Microcosm*, a book that takes as
its theme the "overthrow of matter." On the first page of chapter 1, we
learn that "The powers of mind are everywhere ascendant over the
brute force of things." Though the primacy of the mind over matter is
hardly a new idea in Western philosophy, Gilder writes as if it is. His
universe consists of ideas and the heroic individuals who think them;
his rejectamenta consist of matter and its partisans, the dialectical materialists of the Marxist tradition and the pragmatic materialists of mainstream thought. Society, and with it labor and the state, virtually
disappear from Gilder's view, except in the form of the fickle and ever-
dynamic "market."

And so do class and history. In a classic passage, Gilder erupts:

The United States did not enter the microcosm through the portals of the Ivy League, with Brooks Brothers suits, gentleman Cs, and warbling society wives. [Gilder himself did, however. As Susan Faludi reported in *Backlash*, young George was adopted by the Rockefeller family after his father, David R.'s roommate at Dartmouth, was killed in World War II.] Few people who think they are already in can summon the energies to break in. From immigrants and outcasts, street toughs and science wonks, nerds and boffins, the bearded and the beer-bellied, the tacky and the upright, and sometimes weird, the born again and born yesterday, with Adam's apples bobbing, psyches throbbing, and acne galore, the fraternity of the pizza breakfast, the Ferrari dream, the silicon truth, the midnight modem, and the seventy-hour week, from dirt farms and redneck shanties, trailer parks and Levittowns, in a rainbow parade of all colors and wavelengths, of the hyperneat and the sty high, the crewcut and khaki, the ponytailed and punk, accented from Britain and Madras, from Israel and Malaya, from Paris and Parris Island, from Iowa and Havana, from Brooklyn and Boise and Belgrade and Vienna and Vietnam, from the coarse fanaticism and desperation, ambition and hunger, genius and sweat of the outsider, the downtrodden, the banished, and the bullied come most of the progress in the world and in Silicon Valley.

Somehow, in compiling this Whitmanic catalog, Gilder forgot the teenage women going blind from soldering circuits in the Philippines, the poisoned groundwater in Silicon Valley, the tumors arising in the livers of chip factory workers, the reporters and data-entry clerks paralyzed by repetitive strain injury, and the banalization and cheapening of countless occupations. On a less morbid note, he also forgot the intellectual contributions of researchers in the nonprofit bureaucracies known as universities and in the once-protected monopoly of Bell Labs as well as the decades of government subsidy to the electronics industry via the military budget.

It would be easy to dismiss Gilder as a lone nut, despite his voluminous writings, his posh seats in right-wing think tanks, and the influence of his 1981 book, *Wealth and Poverty,* on the early Reagan admin-istration, an era whose influence is still with us. Not only did the Gipper hand out copies of the book by the boxful; he is even said to have read it, and his chief spook, the now-defunct Bill Casey, subsidized the author during the lean months of composition. Gilder's line is a staple of the business press; the cover of the October 3, 1994, *Fortune,* that slick *Pravda* for the American business class, announces that "Your company's most valuable asset" is "Intellectual capital," which is appar-ently far more important than the physical or monetary kind.

Another reason to take Gilder seriously is that his line on matter's overthrow was also celebrated by Federal Reserve chair Alan Greenspan in an October 24, 1988, op-ed piece in the *Wall Street Journal,* in which Greenspan noted a general trend towards tininess. Chips have replaced vacuum tubes; Thinsulate, fur; and terabytes, paper securities; and intangible, knowledge-dependent services, bulky old-fashioned goods. "In fact, if all the tons of grain, cotton, ore, coal, steel, cement and the like that Americans produce were combined, their aggregate volume would not be much greater on a per capita basis than it was 50 or 75 years ago," he argued with stunning banality. Greenspan's celebration of the immaterial looks especially odd in the light of his youthful faith in the gold standard, one of the most curious of the materialist supersti-tions (this was when he was writing for Ayn Rand's *Objectivist*), and even odder in the light of his present concern with scrap metal prices as a harbinger of future inflation (now that he is our chief money man-darin). In making his argument, Greenspan apparently ignored the evi-dence of his own agency's industrial production indexes, which showed per capita U.S. manufacturing volume up over threefold in the fifty years before he wrote these words, and more than sixfold over the sev-enty-five years.

But it's almost too easy to make fun of Alan Greenspan; back to Gilder and his recuperation of idealism. Though he is a right-wing troglodyte — aside from his celebration of wealth and unfettered capi-talism, Gilder is also a crude celebrant of raw masculine power over women — his cyber-optimism is not unrelated to fashionable post-modernisms in the ideological apparatus, and even to the thinking of liberal and quasi-radical urban planners as well as the centrist Democrats of the Clinton administration.

Pick up a book or essay by a putatively Marxoid urban theorist like Manuel Castells and you read what is essentially Gilder translated from cheerleading journalese into portentous academese — like this passage from Castells' article in *New Left Review* 204:

> By this concept [the informational society], I understand a social structure where the sources of economic productivity, cultural hegemony and political military power depend, fundamentally, on the capacity to retrieve, store, process and generate information and knowledge. Although information and knowledge have been critical for economic accumulation and political power throughout history, it is only under the current technological, social, and cultural parameters that they become directly productive forces. . . . Material production, as well as services, become subordinate to the handling of information. . . .

Information handlers are now accounting for "an ever growing majority" of employment, Castells argues.

That is a line very popular with Robert Reich, a former Harvard professor who now heads Bill Clinton's Labor Department. In his last book as an academic, *The Work of Nations* — which came adorned with an epigraph from Calvin Coolidge and a blurb from George Gilder ("stiletto-sharp"), a fine instance of bipartisanship across time and space — Reich argued that tomorrow belongs to the "symbolic analysts" rather than the "routine producers," to lawyers, bankers, and systems analysts rather than production workers, clerks, and their low- and mid-level supervisors. The key to the American future, in a world where machines and Mexicans are daily replacing the routine producers, is to make more of us into symbol-mongers. Presumably, if the masses were to become conversant with Boolean algebra, our economic future would be secure.

Reich carried this line with him when he moved to Washington. A speech he delivered in honor of the 100th Labor Day, reprinted in the *New York Times* (August 31, 1994), began with a reflection on the "fracturing" of the middle class — continuing his theme of social polarization developed in *The Work of Nations*. He conjured an image of U.S. society divided into three groups, "an underclass largely trapped in center cities, increasingly isolated from the core economy; an overclass, those in a position to ride the waves of change; and in between, the largest group, an anxious class, most of whom hold jobs but are justifiably uneasy about their own standing and fearful for their children's futures."

Aside from the ugly, stigmatizing word "underclass," Reich's picture is a fairly accurate one. A country that once prided itself on its universal middle-classness and the universality of upward mobility — exaggerations to be sure, especially for the nonwhite and nonmale, but with enough of a material base in reality at least to sustain the mythmaking — is now experiencing a profound hollowing out of the middle and a mass experience of downward mobility. What is Reich's solution to this profound social crisis? More training and education. No one could argue with the virtues of learning. But it won't give people jobs. It won't cure polarization. Savor this bit of nonsense from Reich's speech, which would be comical if it weren't so fundamentally callous and out of touch:

> [T]he long-term data also affirm the potential for building a new middle class. Some of the fastest job growth in America is occurring among technicians, who defy the categories of the old economy. In my travels I've met vending-machine repair workers who use hand-held computers to identify problems and communicate with the home office. . . .

This reckless anecdotalism would do Reagan or Bush proud. The future belongs to vending machine repairpersons! If their hand-held computers are designed correctly, they require no great skills — all the skill is programmed into the machine — and promise their operators no high wages.

Is there any truth to Reich's and Castells' blather? How big is the high-tech, infobahn workforce now, and how big is it likely to get? The share of the workforce employed directly in information superhighway kinds of tasks is well under 2 percent — and that includes people who design, make, and program computers, chips, and telecommunications equipment. Business purchases of computer and telecommunications equipment totals just over 2 percent of GDP. What the Bureau of Labor Statistics (BLS — an agency within the department Reich now heads) calls scientists, engineers, and technicians now constitute about 5 percent of the total workforce. By 2005, it reckons, these workers will account for all of 5.6 percent of total employment. Looking at high-tech industries rather than workers gives an even less impressive picture; now they account for just over a quarter of total employment, but by 2005 their share is likely to fall by over a percentage point. Yes, the number of systems analysts and computer scientists will grow dramatically, yes — by almost 80 percent. But since there are under a half-million of such folks now, their share of the workforce will remain nearly invisible to the naked eye. The same can be said of computer programmers, electronics engineers, and biotech scientists.

And what of the future? Let's take a look at the official projections coming from the BLS. Most revealing are the general projections by occupation shown in the table below. A scan down the table doesn't exactly suggest that an information-rich future awaits most workers. Among the fastest-growing job categories will be the clichés of downscaling: sales clerks, cashiers, janitors, security guards. On a fairly generous categorization — generous because there's nothing terribly futuristic about teaching and nursing — jobs with high information content account for 15 percent of total job growth over the next decade, less than half as many as those with low information content.

The 30 Fastest-Growing Occupations, 1992–2005
(ranked by numerical change)

|  | Thousands | Change | Percent of total growth |
|---|---|---|---|
| retail salespersons | 786 | 21% | 3.0% |
| registered nurses* | 765 | 42 | 2.9 |
| cashiers | 670 | 24 | 2.5 |
| office clerks | 654 | 24 | 2.5 |
| truck drivers | 648 | 27 | 2.5 |
| waiters and waitresses | 637 | 36 | 2.4 |
| nursing aides, orderlies, and attendants | 594 | 45 | 2.3 |
| janitors and other cleaners | 548 | 19 | 2.1 |
| food-preparation workers | 524 | 43 | 2.0 |
| systems analysts* | 501 | 110 | 1.9 |
| home health aides | 479 | 138 | 1.8 |
| teachers, secondary school* | 462 | 37 | 1.8 |
| child-care workers | 450 | 66 | 1.7 |
| guards | 408 | 51 | 1.5 |
| marketing and sales supervisors* | 407 | 20 | 1.5 |

| | | | |
|---|---|---|---|
| teacher aides and educational assistants | 381 | 43 | 1.4 |
| general managers and top executives* | 380 | 13 | 1.4 |
| maintenance repairers | 319 | 28 | 1.2 |
| gardeners and groundskeepers | 311 | 35 | 1.2 |
| teachers, elementary* | 311 | 21 | 1.2 |
| food counter and fountain workers | 308 | 20 | 1.2 |
| receptionists and information clerks | 305 | 34 | 1.2 |
| accountants and auditors* | 304 | 32 | 1.2 |
| clerical supervisors and managers* | 301 | 24 | 1.1 |
| cooks, restaurant | 276 | 46 | 1.0 |
| teachers, special ed* | 267 | 74 | 1.0 |
| licensed practical nurses* | 261 | 40 | 1.0 |
| cooks, short order/fast food | 257 | 36 | 1.0 |
| human services workers* | 256 | 136 | 1.0 |
| computer engineers and scientists* | 236 | 112 | 0.9 |
| | | | |
| totals of the 30 shown having: | | | |
| high info content (*) | 3,989 | | 15.1 |
| low info content (all others) | 9,017 | | 34.2 |
| | | | |
| total employment growth (all) | 26,383 | 22% | 100.0% |

*Source:* U.S. Bureau of Labor Statistics figures, reported in George T. Silvestri, "Occupational Employment: Wide Variations in Growth," *Monthly Labor Review,* November 1993, pp. 58–86.

A broader study of the recent past and immediate future by Lawrence Mishel and Ruy Teixeira (*The Myth of the Coming Labor Shortage,* Economic Policy Institute 1991), also based on BLS projections, showed that, as the millennium approaches, likely industrial and occupational shifts will tend to lower, rather than boost, average pay levels. Likely skills demands, Mishel and Teixeira conclude, can be met by a one-quarter of a grade-level increase in the education of new workforce entrants relative to retirees. Further, the most dramatic shifts in skill requirements occurred in the 1960s and 1970s; they've since slowed dramatically, and will probably continue to do so.

But you hardly need a crystal ball or an econometric model to see what the information economy holds; all you need do is walk around New York City, the cutting edge of postindustrialism. While not much hardware is made in New York, the city's leading industries — finance, media, elite business services like lawyering and consulting — are major users of information technology and major contributors of traffic to the fabled information superhighway. For decades, official city policy has been to discourage manufacturing and to encourage the development of the postindustrial information economy. Manhattan's office buildings and universities house more of Reich's symbolic analysts than any other county in the U.S. Lord knows they do well for themselves, very well.

But New York has probably the most bifurcated labor market in the country — heaven for the symbol jugglers, but hell for almost everyone else. As a result, the city's employment/population ratio — the share of the adult population at work, which is a much better measure of work deprivation than the conventional unemployment rate, since it accounts for all those who are too marginalized to be counted in the official labor force — was over 12 percentage points below the national average in 1993. If New Yorkers were employed at the national average, nearly 700,000 more would be working than actually are. This is the future that awaits — not Gilder's world of acne-covered, pizza-break-fasting, Adam's-apple-bobbing geeks.

In this piece, I've focused, unfashionably, on production, mainly because postmodernists and publicists tend to play down its importance and revel in the sphere of consumption instead. I've also focused on the gloomier side of things, for similar reasons: our cyberfuture doesn't want for celebrants. If you strip away the high-tech gloss, this future looks in many ways like the nineteenth century or even the early days of the Industrial Revolution, times of massive polarization and displacement — exciting, even liberating, if you were on the right side of things, but frightful and immiserating otherwise.

I'll close with a return to Castells' astonishing claim that information is now a "directly productive force." Information, no less than physical capital, is meaningless without humans to work it. Like the capitalist apologists who have long sought to make capital an equal factor of production with labor — and to justify, thereby, profit as a reward to production on a par with wages — cyber-apologists are now performing the same trick with information. While information may please or distress, enlighten or confuse, economically speaking information is always information *about* something: a patent that confers competitive advantage, inside knowledge that confers an advantage to the well-placed speculator, or in the Gildered example I opened with — which Castells mysteriously refers to as the "space of flows" — the flow of bytes that merely signify the flows of capital.

Behind all these bits of information are relations of power and often of appropriation. The patent may be of genetic material taken from a Brazilian forest and patented by Merck; the insider's knowledge appropriated from an employer; the capital derived from claims made on real producers and consumers, who service their debts out of profits and wages. But in the celebration of immateriality and heroic individualism these links and dependencies are forgotten.

That forgetfulness is reminiscent of one of Marx's most famous concepts — commodity fetishism. The commodity, he wrote in the first chapter of the first volume of *Capital*, "is a very strange thing, abounding in metaphysical subtleties and theological niceties." It appears as an autonomous, almost self-begotten thing, but in fact it is produced by human labor and realizes its value only in the social process of market exchange. But the market relationship appears to the uncritical eye as merely "a physical relation between physical things. . . . It is nothing but the definite social relation between men themselves which assumes here, for them, the fantastic form of a relation between things." With the info fetish, the thingly relation, and the social relation behind it, appears as the relation between bytes — a second-order fetishism, you might say.

*R. Dennis Hayes*

# DIGITAL PALSY:
# RSI and Restructuring Capital

E very era has its heresies. Perhaps there is no greater contemporary heresy than the notion that computers have betrayed us. Yet betrayal — and fundamental failure — is what the record quietly shows.

The evidence is clear, cumulative, and robust enough to rule out dismissal or denial. And it has been accumulating for over two decades (comparable, in the era of accelerated technology change, to several geological periods). What it points to is this: information technology, upon which we are now very dependent, has occasioned an epidemic of chronic, often crippling, workplace injuries and layoffs. It has also failed remarkably to reduce workloads or, in most cases, to boost productivity. As a result, computers have significantly raised the cost of doing business and intensified the pace of work without delivering on many of the benefits presumed, promised, or imagined to have accompanied the information age.

How shall we reckon the unexamined impacts of computers on the U.S. workplace? Let us call them, for obvious reasons, countertrends. Two of them are graphically depicted here alongside the mainstream trend of computer sales.

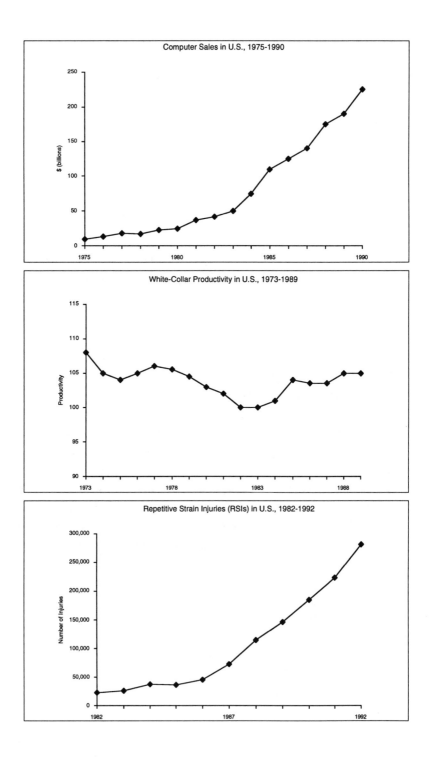

Computer Sales in U.S., 1975-1990

White-Collar Productivity in U.S., 1973-1989

Repetitive Strain Injuries (RSIs) in U.S., 1982-1992

## Computer Sales

The curve representing U.S. computer sales tracks more than two decades in which firms large and small spent over $1 trillion on information technology (Myron Magnet, "The Productivity Payoff," *Fortune,* June 27, 1994). Dollar for dollar, U.S. industry invested more in information technology during the last four years than in all other kinds of capital equipment combined, including all the (noncomputer) machinery needed for services, manufacturing, mining, agriculture, and construction. The end of the twentieth century will witness more of the same, thanks in part to a bipartisan rollback of antitrust laws (particularly in the banking industry) and to new or proposed technology policies. Computers, it seems fair to say, are now a necessary and growing cost of business.

## White-Collar Productivity

The curve depicting productivity is from a study of information workers, who, by 1994, accounted for over half of all U.S. employees. This countertrend — the tendency of productivity rates to sag or flatten at workplaces investing in information technology — is, understandably, widely unreported. But it has attracted the attention of more and more economists who call it the *productivity paradox.*

Productivity — a kind of "rate of exploitation" — gauges workplace efficiency from the proprietary view concerned with generating capital and profit. In healthy industrial economies, it is expected to rise each year. And in the post–World War II period it did, growing steadily until the mid-1970s. This is roughly the time when investments in information technology began to take off. While rising still in the U.S. manufacturing industries (the rise is not adjusted for hidden costs), measured productivity growth has fallen sharply in the so-called service industries, which is where the U.S. has invested over 80 percent of all information technology (Erik Brynjolfsson, "The Productivity Paradox of Information Technology," *Communications of the ACM,* December 1993).

The graph shows an upward trend, beginning around 1984–85, in service industry productivity. This reflects the impact of downsizing and subsequent speedups that accompanied restructuring in the American workplace in the 1980s and that continue to this day.

Companies restructure by laying off workers and managers and by redesigning jobs around information technology. Where productivity is rising at all, it is mainly in industries that have computerized job tasks or laid off workers (often to offset the cost of computerization) or both. But the introduction or expansion of computers has not resulted in a net decrease in workload per worker. As a result, most of us are doing more work, and many of us are getting injured.

The productivity curve does not reflect the full burden of computers. No one knows (because no one measures) how much is spent on litigating, treating, retraining, and replacing the computer-injured. In 1992 a physicians group hazarded a guess: $25 billion per year (Mark Pinsky, *The Carpal Tunnel Syndrome Book*, New York: Warner, 1993, 17).

### Repetitive Strain Injuries

Workers who use computers, or whose jobs have been redesigned to accommodate computers, are suffering historically significant rates of workplace injuries. What doctors and health insurance firms call repetitive strain injuries (RSIs) and cumulative trauma disorders (CTDs) are now the most commonly reported workplace illness (by far — nearly 63 percent as of 1992) (*VDT News*, January/February 1994).

RSIs are unlike other known or suspected computer-related health threats, which range from eyestrain and skin rash to radiation exposure and miscarriage. RSIs stem from awkward, forceful, or rapid hand and wrist movements that traumatize and damage the muscles, tendons, or nerves of the neck, shoulder, forearm, wrist, or hand. The injuries, which typically occur in the stressful context of unrelieved work, are chronic. And the symptoms include strong, recurring pain as well as weakness, numbness, and motor impairment.

The sharp rise in recent rates — an increase of nearly 1,246 percent from 1982 to 1992 — parallels the ascendance of information technology in corporate restructuring. It also invites comparisons to the first waves of black lung, brown lung, and asbestosis. Although nonfatal, RSIs can be just as disabling as these earlier workplace-contracted afflictions. And RSIs are potentially far more widespread. About half the U.S. working population is at risk. By the year 2000, the at-risk figure will rise to 75 percent of all workers (Emil Pascarelli and Deborah

Quilter, *Repetitive Strain Injury: A Computer User's Guide*, New York: Wiley, 1994, 3). Still, the incidence and nature of computer injuries remains grossly under-reported and deeply misunderstood.

The Bureau of Labor Statistics (BLS) did not survey public employees, including postal workers, whose susceptibility to computer injuries is notorious. And "white-collar" workers were underrepresented in the RSI surveys, even though the trade, financial, and service industries accounted for nearly all the increase in reported RSIs in 1992.

Nor are BLS methods accurate, according to an open letter sent by a coalition of unions in 1991 to then–Secretary of Labor Lynn Martin. The coalition called for an emergency safety standard to address what they called an "epidemic" of RSIs stemming from "real and substantial" changes in the restructured American workplace. Contending that the BLS undercounted the incidence of RSIs, the coalition cited research showing that "rates of work-related cumulative trauma disorders are actually 50 to 100 times greater than rates reported" by the BLS. That would augment the 1991 BLS estimate of 223,600 injuries to a staggering 11 million to 22 million (Pinsky, 145).

Do computers cause repetitive stress/cumulative trauma ailments? A comparable set of symptoms has been linked to work for at least *8,500* years. The skeletons of neolithic farm women show grossly arthritic big toes (a condition developed while kneeling to grind cereals on a saddle-shaped stone mill) (Norman Hammond, "RSI Traced Back to the Stone Age," *San Francisco Examiner*, August 12, 1994). In mid-eighteenth-century Italy, a researcher describes a "disease of clerks and scribes" from "continuous sitting, repeated use of the hand and strain of the mind." A nineteenth-century British medical journal recorded similar conditions among milkmaids and shoemakers. And at the turn of the century, roughly 60 percent of Britain's telegraph operators suffered repetitive strain injuries (Cited in Edward Felsenthal, "An Epidemic or a Fad?" *The Wall Street Journal*, July 14, 1994).

The relationship between RSIs and information technology is a special one. The primacy of computer technology in the workplace has been a great leveler, affecting managers and managed, factory and office workers. By combining jobs, clustering work tasks, and monitoring performance, business firms have used information technology to radically revise the way nearly every employee works. The problem is that computers, more than any single previous technology, have funneled work tasks into a very narrow range of physical motion. Add to this the prod of ever more rapid computer hardware and software as well as the overtime induced by downsizing: layoffs and budget shortfalls. The subsequent greater pace and stress — in a word, *intensity* — of work has transformed workplace and worker. As a NIOSH researcher wrote in 1988 (Vern Putz-Anderson, cited in Pinsky, 12–13):

> Automation has been successful in shifting the locus of work from the level of the trunk to the upper extremities [arms]. The workloads are now lighter, but the workpace has been increased. As a result, the associated work forces are concentrated on smaller parts of the anatomy, i.e., the ligaments, tendons, muscles, and nerves that control the hands, wrists, and arms of a worker.

The plight of the computer injured is bleak because of (1) misdiagnosis by doctors, by employers, as well as by workers' compensation, job disability, and insurance review panels; (2) the real, if hidden, cost to workers for whom RSIs typically are chronic and often disabling; and (3) the reluctance by the injured party, fearing stigma, isolation, and job loss, to report the condition, and by the employer, fearing liability, to acknowledge the dimensions of the problem.

Computer design and work pace are far from immutable. In fact, they are utterly negotiable. Both currently reflect the prerogatives of generating capital and controlling the networks and hierarchies of work.

At every level, the design of information technology is dominated by a Newtonian quest for speed and raw computing power that dovetails with the corporate demand for faster transaction processing. Speed is perceived as competitive advantage. The presumption is that more work can be performed by a faster machine than by a slower machine. This is true enough, mechanically speaking. But this design principle completely ignores the effects on the human operator. More to the point, it imagines an operator of superhuman resilience.

Faster machines beckon us to work faster, a proposition that is viewed favorably by employers. But where the limits to machine speed preoccupy design engineers, machine speeds that encourage a prudent and safe pace of work are ignored. This is why remedies for RSIs based on litigation and ergonomic redesign are likely to fall short. Hope for a solution based on government intervention (OSHA/NIOSH) or on prevention based on enlightenment and mutual (worker/employer) compliance is likewise misplaced. The government's policy on computers is nothing short of promotional.

A cavalcade of therapies and treatments, nearly all of them (necessarily) experimental, have emerged in recent years. They range from surgery to calisthenics to ergonomically designed office furniture. Some of them seem to work some of the time. What can be said is that an effective and lasting intervention awaits a slower pace of work and a workplace redesigned around convivial computer tools. Make no mistake: in 1995, these are almost revolutionary projects.

In the meantime, the sheer number of RSIs has already prompted recognition. For the record, computer manufacturers and employers deny culpability for RSIs. To limit their liability, many of them install flexible office desks and do not block workers' compensation checks. This year, two giants in the computer industry (Compaq and Microsoft) included RSI warnings on their keyboards. Unfortunately, without official acknowledgement, a kind of triage has emerged whereby injured managers and salaried professionals quietly receive funding for treatment and retraining while similarly afflicted retail clerks and data-entry operators even more quietly get less or no consideration. It will take a struggle by everyone at every level to win an acknowledgment of consequence. This is particularly true in the U.S., where, to cite just one example, cotton dust was disavowed as a workplace hazard in the textile industry for forty years after it was a treatable condition in Great Britain.

·

The countertrends point to a deep irony of the information age: how misinformed we are about information technology's real-world effects and, in particular, about its threats to our personal health and economic well-being. The countertrends also raise disturbing new questions about information technology. Our answers need not be constrained by predilections to turn back the clock or to make the best of what seems inevitable.

The countertrends represent existing phenomena that allow of exceptions and some interpretation. But taken together they challenge some of the most strongly felt views in the U.S., including those informing official government business, technology, and job policies.

Corporate, academic, technical, and political leaders hail information technology. They invite us to get it, learn how to use it, and build newer and faster versions of it as quickly as we can. A new conventional wisdom gauges progress by the number of computers per workplace and classroom, the speed and size of the latest gadget, the special effects of the season's most successful action film. A Malthusian school of business analysis applauds layoffs and computerization, the more the better, as a step toward economic health through restructuring. It can't be said enough: companies restructure by laying off workers and managers and by using information technology to combine jobs, to concentrate and computerize work tasks, and to increase employee workloads. Is it really surprising that so many have suffered injury?

*Monty Neill*

# COMPUTERS, THINKING, AND SCHOOLS IN THE "NEW WORLD ECONOMIC ORDER"

> Capitalism is the first productive system where the children
> of the exploited are disciplined and educated in institutions
> organized by the ruling class.
> — Mariarosa Dalla Costa and Selma James

Two primary reasons are given for plugging schools into the National Information Infrastructure (NII; called "the information superhighway"). Plugging-in is required to prepare students to be highly skilled, highly paid workers in the economy of the future; and it is essential for reforming schools into institutions that will produce students who can think and solve problems.

The "double helix" of high-skill jobs and cognitively complex schooling is presented as liberatory (Berryman and Bailey 1992). But the use of computerization toward a distinctly nonliberatory end is the more likely consequence of the twinning of school and work in the emerging world capitalist economy.

## Computers and the Jobs of the Future

Both conservatives and liberals argue that for the U.S. to "be number 1" in the world economy a more educated working class is needed, one that works harder *and* smarter. The claim is that then corporations will create jobs that utilize the workers' skills. These high-skill workers will be more productive than others and will therefore earn high wages. The alternative, they warn, is low skills and low wages. Schools must therefore educate "all" students to "world-class standards" so that the corporations will be competitive.

This argument is made in the report *America's Choice* by the National Center on Education and the Economy (NCEE 1990), probably the most influential piece on U.S. education since *A Nation at Risk* (fraudulently) maintained that falling school quality endangered national security. The NCEE view can be found in legislation, particularly the recently enacted Goals 2000 school reform bill; in numerous corporate education reform proposals (e.g., California Business Roundtable 1994); and various books and government reports (e.g., Carnevale and Porro 1994). It has become virtually unquestioned conventional wisdom.

The most obvious thing about this claim is that it presumes an uncontrollable and inevitable economy to which "we" must adapt. It demands that educators accept, not challenge — never mind reconstruct — the economy.

Yet two points about the emerging economy suggest that "we" should not accept it. First, continual lowering of wages is already fact and not likely to turn around; second, most new hires are not likely to be doing high-skill work. For U.S. workers, real wages have been declining nearly 1 percent per year for two decades, while the dispersion of the wage — the gap between high- and low-wage work — has simultaneously widened.

As Midnight Notes (1992) argues, this calculated intent of the capitalist system over the past twenty years to reduce working-class power and income around the globe has had substantial success. Its political and technical ability to move products and services rapidly around the world has weakened the capacity of working people to band together at the national level to push up or even maintain wages. Since the competition for jobs cannot be contained by national borders, wages are dropping toward the lowest levels among the competitors, even for many high-skill jobs such as computer programmer. This push is, if anything, intensifying. The North American Free Trade Agreement, for example, is organized as a one-way ratchet to continue the lowering of wages in Mexico, Canada, and the U.S. (Calvert and Kuehn 1993), to intensify what Kuehn terms "the race to the bottom." There is certainly no reason to believe that the capitalist system will create a worldwide high-wage system or that the U.S. will remain immune from wages within its borders falling to "world-class standards."

The fallacy that most jobs will be high-skill is also widely accepted. Yet even strong proponents of the claim, such as Bailey (1991), acknowledge that most new hires for at least a decade will be filling old slots that do not require the knowledge and skills that proponents of the "high-skills" argument point to. Moreover, the labor market forecasts that project growth in the U.S. in medium- or high-skill jobs do not consider changes in the world economy that are dispersing skilled employment more widely while driving down wages. At most, the number of middle-level-skill jobs will grow slightly in the coming decade.

Even school reformers whose first interest is not in creating workers to serve the economy nonetheless buttress their reform proposals by pointing to the presumed high-skill information economy. But what are the implications of all this for schools? Lower wages coupled with continuous attacks on public services and increased class and race stratification — the actually existing U.S. conditions — strongly suggest continuation of the "savage inequalities" so eloquently described by Jonathan Kozol (1991): sharp class gradations with immiseration for many.

The way computers and paraphernalia have been distributed already indicates this (Piller 1992; Pearlman 1994; Ramirez and Bell 1994; SEDLetter 1993). Not only are rich kids more likely to have computers at school, but their schools' machines are more apt to be up to date, drive more sophisticated software, and be connected to the Internet. Most schools are barely wired — most classrooms don't have telephone jacks or the electric wiring to run more than a couple of computers. The less money a school or school district has, the less likely it will be able to ride on the information superhighway. Presuming that funding can be found for these essentials, schools still must raise money to stay on line and educate teachers in technology use: over a five-year period, the hardware is only 18 percent of the cost of using technology (Van Horn 1994).

Telecommunications corporations are eager to exploit the school market, sometimes offering to wire schools in exchange for controlling the wires that will hook the schools to the Internet and thus to corporate coffers (Coile 1994; Einstein 1994). Poor schools in particular are prey for technology profiteers such as Whittle Communications' now-defunct Channel One. Channel One provided satellite dishes, VCRs, and TV monitors to schools that agreed to force students to watch a ten-minute daily "newscast" that included two minutes of ads. Schools in poor neighborhoods or those with the lowest per student annual spending were respectively two or six times more likely to have Channel One than were schools with the wealthiest students or highest per student spending (Morgan 1993).

Health and safety issues of computer use are also most stark for poor schools. Children are more susceptible to radiation, including that from computers, and are also at risk for the same muscular and eye strains as adults (Miller 1992). Schools that can barely afford computers are least likely to shield them or purchase ergonomically correct furniture.

In sum, the savage inequalities of the past will extend into the wired savagery of the future. There is neither empirical nor theoretical reason to believe this scenario will change for the better so long as the capitalist system continues. In general, students from low-wage families and communities need more resources if they are to catch up in the kinds of skills (technical, academic, and cultural) sought for in high-wage occupations — yet they get substantially less. Why expect the capitalist system and its government to invest extra funds to develop low-income children into sophisticated problem-solving workers if the jobs don't and won't exist?

In any event, wherever the system does invest in schooling, its purpose is to intensify schoolwork by children and to prepare them for future work — while presenting this as in the students' interest. The call for students to work harder in school is as ubiquitous as the call for schools to produce high-skill workers — and usually comes from the same sources.

Computerization of schools will not contribute to "high wages" or "good jobs." The U.S. class hierarchy will not be ameliorated by computerization, but will be intensified. Indeed, computers have been a fundamental weapon in the capitalist war against the working class over the past two decades. If computer knowledge is required in the economy, it is not because of any capitalist desire for highly paid workers or any great need for highly skilled workers.

## Thinking Machines for Thinking Students?

A strong claim is sometimes made that using computers and related high-tech machinery is *necessary* for a change to a mode of schooling that focuses on thinking. For example, Ramirez and Bell (1994) conclude

> It is the position of this paper that if systemic school reform in this country is to succeed it will only do so with the application of telecommunications and information technologies at the classroom level with a simultaneous focus on sustained professional development for teachers.

The argument rests on the purported necessity of computers for enabling all students to engage in higher-order thinking activities such as understanding complex ideas, solving real-world problems, and analyzing critically.

There is an irony in this claim. The emphasis on higher-order thinking in schools rests substantially on the foundation of cognitive psychology. As Noble (1989) has shown, cognitive psychology evolved in large part because the U.S. military wanted to create artificial intelligence — but it had no useful understanding of the genuine thing. The military therefore funded extensive research in cognition, research that ended up largely confirming what progressive educators and psychologists had long maintained, that humans learn actively and by constructing and modifying mental models.

The dominant psychological theories in the U.S. have been behaviorist. In the version influencing schooling, humans supposedly learned by passively accumulating isolated bits of information. In time, the bits could be shaped into successively more complex patterns. The impact, however, was that schools presumed students could not think in a given area until they had accumulated enough bits. Those who did not sufficiently grasp the bits were condemned never to do anything interesting in school. This approach still dominates curriculum, instruction, and the ubiquitous standardized tests. Cognitive psychology, however, proposes learning as a fundamentally different process. Recognizing that students think, learn by thinking, and can learn to think better or differently, it calls for a "thinking curriculum" (cf. Resnick 1987).

The irony here is that having constructed cognitive psychology in order to develop "thinking machines," the machines are now presumed indispensable for helping students learn to think. But the very existence of thinking in schools without computers shows clearly that machines are not necessary for a thinking curriculum. The reason schools haven't encouraged thinking is not because they have lacked computers, but because the system did not want thinking workers.

A softer claim for the necessity of computers is that because of the way the economy relies on computerization, only via computers will it provide access to materials and knowledge that will facilitate higher-order thinking in academic areas for many more children. Presumably, the NII will enable access to teachers and learned people, data banks and libraries, analytical tools such as statistical packages, and other software.

Despite the absence of funds, it is claimed that teachers working in poor systems will be able to get this complex operation functioning. However, if teachers had the time, training, and support to do this, they could reorganize their classrooms for inquiry, dialogue, critical thinking, understanding, and problem-solving — with or without computers. They don't succeed for a number of reasons: too many students, lack of resources and knowledge, and standardized tests that militate against thinking. Somehow, though, the computer will be the means to make the instructional leap.

Still, the claim of the value of computers has some persuasive aspects, though with many caveats:

• The tools free up time. For example, because of calculators, rather than spend time on arithmetical drill, students can spend time on learning mathematical reasoning and problem-solving.

• Access to information is enhanced. In many schools, the library is outdated or inadequate. The cost of some information will cheapen, making it more accessible — assuming that poor schools have the money to get and stay on line. Access, however, says nothing about the nature of the materials available on line. In seeking information, whether on networks or elsewhere, one is limited by one's inquiry framework. Without a strong frame, a student will simply be buried in tons of data. Access to information means little without guidance in learning to use information — which raises questions of whose guidance for what purposes.

• Access to people is expanded. You can use the Internet to dialogue with people all over the planet. (Of course people from all over the planet may now live in your neighborhood.) Computer advocates constantly tout examples of "real" scientists talking with kids from some school, but once millions of students are on line, how many scientists will spend time sorting through hundreds of on-line requests?

• Access to some kinds of computerized tools enable students to work on sophisticated problems rather than more basic and boring ones. Working on more realistic and complex tasks, doing so in collaboration with others, proceeding at one's own pace, even having a real-world use for the results, all can help motivate students. Again, much of this does not require computers or the NII. Moreover, once the technology becomes old hat, deeper issues of the purpose for schooling will inevitably reappear for students, for whom lack of control often guarantees lack of interest (Herndon 1972).

The important issues are not ones of technology but of politics: will the funding be there? what kinds of guidance to what ends will students receive? who controls the technology? for whom will the computer ultimately be useful? Class relations that are played out in technology implementation are also implicated in technology construction. What is done with tools is not determined so much by those who use the tools as by those who construct them. Thus, the ways in which the makers design technology can largely control the structuring and solution of problems by users, to whom the control by the maker remains invisible (Madaus 1993). Computer use is then falsely promoted as a neutral yet liberatory tool.

## Controlling School and Work

The controlling class no more wants problem-solvers and critical thinkers to do most jobs of the future than it did in the past, during the assembly-line era. Computer use in schools *will* fit the economy — not the mythi-cal economy of "high skills and high wages," but the real economy of "the race to the bottom." While following orders, not questioning, being on time, and submitting one's personality to the dictates of the school all pre-pared workers for jobs in the mass-production era, the schoolwork form directly fit the actual jobs for only a relatively few workers. With comput-erization, however, form can more closely resemble function.

The McDonald's level of familiarity with technology requires no actual knowledge of computers or much thought. Data-entry (with the computer monitoring your speed) and similar work does not require higher-order thinking. Schools will train students to sit in front of com-puters and do routine work in direct preparation for their jobs. For them, this will be their real-world learning connection.

Use of computers at the technician level sometimes does require decision making, but the parameters are usually specified carefully, meaning that the thinking done is not at the order of making definition but of application. (A look at the actual jobs described by Bailey (1990) or Zuboff (1988) reveals this.) These jobs do require more academic — school-based — knowledge and the ability to apply that knowledge, and the number of these kinds of medium-level-skill jobs probably will increase in the coming decade. Controlling the development and nature of the thinking of those who have limited-problem-solving jobs will be another task of schools.

Data-entry, monitoring, and limited problem-solving will contin-ue to comprise most computer use by the great majority of employees in the U.S., barring an upheaval against the jobs system. Noble's (1991) critique of Zuboff points out that the "intellective" work she glorifies in fact includes two kinds of work that imply a corresponding schooling: one that is scientific and problem-solving and another that is primarily monitoring the process — a difference that "reflects a cavernous hier-archical division of labor." And, Noble adds, the latter is more "about attitudes and disposition than about 'knowledge' or intellectual abili-ties." The mind is reduced to the hand. For most, schooling, wired or not, is preparation for routine work, same now as it ever was. However, I suspect that though "the more things change" is still operative, there are yet some important changes in the offing. The changes have to do with capitalist control over thinking.

As Noble (1989) explains, behaviorism largely treated thinking as a black box. On the assembly line, it did not matter what, or if, the worker thought, as long as he or she behaved: came to work, did the job, didn't cause trouble. To the extent that thinking was an issue, the concern was how to manipulate the worker into working harder (i.e., to control behavior). Industrial psychology developed as a tool to help organize and ensure the functioning of the productive process. It developed a knowledge base that retains its usefulness for management, because traditional "good worker" characteristics are still those most desired by the bosses (NCEE 1990).

In a system that did not want workers to think too much but needed to control their actions, behaviorist psychology was useful. Thus, corporations, foundations, and government agencies funded research that provided ready tools for shaping schooling and controlling workers.

Cognitive psychology is more useful to today's system, which needs workers to think for the system and to think differently, manipulating abstract symbols. The schools are to provide these skills. Those who use computers to analyze, create, or control will be few in number but important to the system (Bailey 1991; Reich 1992). They too must be programmed, but with allowance for a greater degree of self-regulation.

The danger of progressive education that expects students to think and problem-solve is that it might get out of control, leading students to "unrealistic expectations" and a command of areas of knowledge useful for attacking the system. But as Dalla Costa and James (1975) note, as long as progressive schooling remains within control, it not only presents no danger, it may be a source of yet greater profits (see also Robins and Webster 1989, 218–25). The question for the owners of capital, then, is how to ensure control of the "thinking" curriculum. Revealingly, much of the proposed school reform that rests on cognitive science as the model for instruction still rests on behaviorism for motivation and discipline.

The plans of the New Standards Project are perhaps most illustrative. (The founders of New Standards are leading cognitive psychologist Lauren Resnick and Marc Tucker, head of the NCEE, which produced *America's Choice,* calling for "high skills and high wages.") New Standards, which has signed up sixteen states in its development program, proposes not only a new curriculum, instructional methods, assessments, and professional development for teachers, but also the use of performance levels and tests to measure student progress toward goals that are substantially about workforce preparation. Nonpromotion and nongraduation are the negative reinforcers to complement the presumably more interesting new curriculum.

So what's new? Surely not the drive for control or the use of behaviorism. What is potentially new are the means of control, the computer itself, and the target of control: thinking. On one level this is already quite visible in the use of the work tool to monitor the pace of the work. In schools, however, the issue is more subtle. For example, the application of computers to real-world problems will teach students how to solve problems on terms amenable to the controllers of the system and will sort out those who are most willing and able to do so. The "less able" will be funneled to less cognitively complex computer work to prepare them for lower-skill jobs (or lack thereof), while the less willing will be driven out.

The determination of "less able" is a matter of assessment. While assessment is a necessary part of learning, it is all too likely that emerging cognitive techniques will simply become a more sophisticated method for sorting students. Much of this may be done via computerized exams, enabling a high degree of standardization to "world-class" levels; currently, the same old tests used for sorting by class, race, and gender are being adapted for computer (FairTest 1992). Down the road, this could entail use of sophisticated means of analyzing everything from problem-solving ability to personality constructs to degrees of willingness to work (Raven 1991). This knowledge is used, however, not to help students but to control them (Robins and Webster 1989).

Moreover, the computer itself will be used to shape the personality. The model is the computer — the malleable, controllable, programmable "smart machine." Part of the information-technology agenda is to learn how better to control the thinking of humans. At the crudest level, schools will try to do what they have always tried to do, shape students into workers, but the more subtle strategy is to make the mind *want* to be computerized. Perhaps the child must be caught at a young enough age so that she is less able to resist effectively.

Thinking is redefined as what computers do or what humans do to interact with computers, eliminating the rest of the mind and body from thinking. Zuboff (1988) explains how paper workers historically used smell, touch, and direct sight on the job, and she understands this as a widely generalized use of intelligence, an intelligence destined to be replaced by more abstract modes, tied to symbols on computer screens. Thus, the alienation of the body, long a trend under capitalism (Midnight Notes 1982), leaps to a new qualitative level as the definition of thinking is reshaped to meet new capitalist needs.

Social alienation also intensifies as humans interact via the computer, a form of interaction virtually stripped of emotional and social cues. Already some hints are emerging that extended replacement of in-person interaction by virtual interaction decreases a person's ability to socialize comfortably with other people when in their physical presence. Programs in which students work collaboratively on computer projects will ameliorate this tendency, but the students will still be learning "skills" needed to desocialize themselves. Privatization of schooling will further desocialization because it will increasingly allow schooling at home; contact with others will be only via the wire. (Michigan State has already awarded "charter school" status, and thus funding, to a "school" that is basically an electronic hookup among home-schoolers; the "school" is organized around "Christian" fundamentalist ideology (Walsh 1994)).

In inducing physical and social isolation, the computer is the extension of the "white man." Devoid of emotion, disconnected from the body (except during a *work*out), nonnurturing and unmusical, the type of the "white man" excludes all the human traits capitalism has attached to women and people of color (particularly Africans). I am not talking genetics but about the historical hierarchical division of labor that associates human qualities with the work one is forced to do and then calculatedly reproduces those qualities in people in order to force them to do the work (James 1975). The "white man" — really, *bourgeois* — qualities are now to be extrapolated and intensified, abstracted into the computer and then used to school the child into being computerlike.

Complex problem-solving can itself be a form of mindlessness. The language of school reform pays some attention to the issue of "habits of mind." On the one side, it suggests that students learn to think about their thinking, learn not to just accept but to probe, question, challenge. Well and good if it happens, but it is more likely that "habits of mind" — that is, "critical thinking" — will be confined to areas defined and controlled by the system.

The computer thus will be used to control how one learns to think in order to subsequently control how one thinks. The trajectory of the capitalist use of computers goes from constructing cognitive science for planning artificial intelligence to using artificial intelligence to control thinking as defined through cognitive science. Salomon, Perkins, and Globerson (1991) argue that use of intelligent technologies will make humans themselves more intelligent, provided, however, that these advances are "cultivated through the appropriate design of technologies and their cultural surrounds." The definitions of intelligence, the uses to which intelligence will be put, who is to be made more intelligent and how, and the questions of purpose, design, and control are deferred for "people of different expertise" — academicians all — to debate and plan.

The idea that the capitalist system wants a good many critical thinkers is simply absurd — it can only spell trouble unless the thinkers are thinking for, not against, the boss. Thus the point is to produce the human as puzzle-solver, not really as critical thinker. Puzzles can be entertaining, challenging, require lots of thought, and yet be substantively mindless. The mind is thus habituated to thinking only in limited, even if complex, ways.

The beauty of the computer is not simply the speed with which it computes, nor even all the troublesome work-resistant workers it can replace, but that it can simultaneously powerfully shape the mind and the personality. Thus, if *successful,* computerization will enable the production of the human as computer.

·

The world has for centuries been dominated by capital, with its exploitation of a hierarchy of labor power starting with the unwaged (Dalla Costa and James 1975) and interwoven with factors of gender and race. While working-class resistance has pushed the capitalist system to crisis, the working class has not resolved the crisis in its favor (Midnight Notes 1992). Now, as capitalism appears on the advance and recomposes the working class — while incorporating all its old forms, from slavery on — we see the spread of fantastical illusions, from fundamentalist religion to computers, as liberation from the miseries of the world. More mundanely, the use of computers in schools is presented as opening up the possibility of schooling as exciting, powerful learning that leads to "better jobs." Yet the excitement and power will not be for the many — unless the many change society, its economy, politics, and social relations, and how it educates its children.

Only egalitarian, collective, working-class power can assure that any particular technology will be used for its benefit — if it should be used at all. Liberation is not a matter of technology but of social relations. A first step in the transformation of social relations is to refuse the inevitability of the economy.

*Acknowledgements*
Thanks to Shelley Neill, Bob Schaeffer, and Midnight Noters for insights, comments, and support.

*References*
Bailey, Thomas. 1990. "Jobs of the Future and the Skills They Will Require: New Thinking on an Old Debate." *American Educator* 14 (1).

———. 1991. "Jobs of the Future and the Education They Will Require: Evidence from Occupational Forecasts." *Educational Researcher* 20 (2).

Berryman, Sue E., and Thomas R. Bailey. 1992. *The Double Helix of Education and the Economy.* New York: Institute on Education and the Economy, Teachers College, Columbia University.

California Business Roundtable. 1994. *Mobilizing for Competitiveness: Linking Education and Training to Jobs.* San Francisco: CBR.

Calvert, John, and Keuhn, Larry. 1993. *Pandora's Box: Corporate Power, Free Trade, and Canadian Education.* Toronto: Our Schools/Our Selves Education Foundation.

Carnevale, Anthony P., and Jeffrey D. Porro. 1994. *Quality Education: School Reform for the New American Economy.* Washington: Office of Educational Research and Improvement, U.S. Department of Education.

Coile, Zachary. 1994. "'Free' Computer Revolution Now Has a Price Tag." *San Francisco Examiner* (January 30).

Dalla Costa, Mariarosa, and Selma. James. 1975. *The Power of Women and the Subversion of the Community.* 3d ed. Bristol, England: Falling Wall.

Einstein, David. 1994. "Pac Bell to Wire State's Schools for High Tech." *San Francisco Chronicle* (February 19).

FairTest. 1992. *Computerized Testing: More Questions than Answers.* Cambridge, Mass.: FairTest.

Herndon, James. 1972. *How to Survive in Your Native Land.* New York: Bantam.

James, Selma. 1975. *Race, Sex, and Class.* Bristol, England: Falling Wall.

Keuhn, Larry. 1994. "NAFTA and the Future of Education." Paper for the annual conference of the National Coalition of Education Activists (August). Portland, Oreg.

Kozol, Jonathan. 1991. *Savage Inequalities: Children in America's Schools.* New York: Crown.

Madaus, George. 1993. "A National Testing System: Manna from Above? An Historical/Technological Perspective." *Educational Assessment* 1 (1).

Miller, Norma L. 1992. "Are Computers Dangerous to Our Children's Health?" *PTA Today* (April).

Midnight Notes. 1982. "Mormons in Space." *Computer State Notes.* Boston: Midnight Notes.

———. 1992. *Midnight Oil.* New York: Autonomedia.

Morgan, Michael. 1993. "Channel One in the Public Schools: Widening the Gap." A research report prepared for UNPLUG. Amherst, Mass.: Author.

National Center on Education and the Economy (NCEE). 1990. *America's Choice: High Skills or Low Wages!* Rochester, N.Y.: NCEE.

The New Standards Project. 1992. *A Proposal.* Pittsburgh: Learning Research and Development Center & National Center on Education and the Economy.

Noble, Douglas D. 1989. "Mental Materiel: The Militarization of Learning and Intelligence in U.S. Education." In *Cyborg Worlds.* Edited by Les Levidow and Kevin Robins. London: Free Association.

———. 1991. "In the Cage with the Smart Machine." *Science as Culture* 10.

Pearlman, Robert. 1994. "Can K–12 Education Drive on the Information Highway?" *Education Week* (May 25).

Piller, Charles. 1992. "Separate Realities." *Macworld* (September).

Ramirez, Rafael, and Rosemary Bell. 1994. *Byting Back: Policies to Support the Use of Technology in Education.* Oak Brook, Ill.: North Central Regional Educational Laboratory.

Raven, John. 1991. *The Tragic Illusion: Educational Testing.* Unionville, N.Y.: Trillium.

Reich, Robert B. 1992. *The Work of Nations.* New York: Vintage.

Resnick, Lauren B. 1987. *Education and Learning to Think.* Washington: National Academy Press.

Robins, Kevin, and Frank Webster. 1989. *The Technical Fix: Education, Computers, and Industry.* New York: St. Martin's.

Salomon, Gavriel, David N. Perkins, and Tamar Globerson. 1991. "Partners in Cognition: Extending Human Intelligence with Intelligent Technologies." *Educational Researcher* 20 (3).

SEDLetter. 1993. "Cyberschooling: Fiber Optic Vision? Or Virtual Reality?" *Southwest Educational Development Laboratory Newsletter* 6 (1).

Van Horn, Royal. 1994. "Building High Tech Schools." *Phi Delta Kappan* (September).

Walsh, Mark. 1994. "Charter School Opponents Taking Cases to Court." *Education Week* (October 7).

Zuboff, Shoshana. 1988. *In the Age of the Smart Machine: the Future of Work and Power.* New York: Basic Books.

*Daniel Harris*

# THE AESTHETIC OF THE COMPUTER

Although the appearance of our appliances is largely determined by
their use, even the most prosaic of utensils have what might be
called an "aesthetic," a look, a style that serves no discernible function
and that reflects the sensibility of their designer. Something as insignificant as a spoon, an implement in which utility clearly predominates
over aesthetics, is stamped with a floral pattern, just as cheap toasters
have textured plastic panels that simulate leather or upholstery, or
refrigerators are embossed with metallic snowflakes, or colanders have
clusters of holes shaped like flowers — features that affect the user in a
less tangible, more poetic way than the pure utility of the notches on
a steering wheel, the plastic grip of a can opener, or the sandpaper
surface of a Daisy-Mate bath decal.

An appliance as complex as a computer also has gratuitous, decorative elements that serve only to beautify the machine. These fanciful
embellishments can be found, not only on the external surfaces of the
hardware, but in more elusive and internal places within the software
itself. Here, in these hidden recesses, the user communes with the
machine as he works, engaging in a private conversation that consists
entirely of images, some of which are of instrumental importance to
the job at hand, while others function solely to raise his spirits with
the jokes, quips, and gags that are now standard components of many

computer programs. One of the most popular software packages on the market, Windows, is cluttered with comic-book froufrou, with frills and trimmings that fill the backgrounds and margins of the screen like manuscript illuminations — the visual pun of a spectral kitten, for instance, that leaps around on the screen after the "mouse," following the cursor wherever it goes and pouncing on it when it stops moving. With the help of such ornate details, which software designers now scrawl over their programs like graffiti, the computer is most decidedly acquiring an "aesthetic," a pictorial style that, while serving no conscious, utilitarian function, serves a number of *un*conscious, psychological ones.

Nowhere in the recent revolution in the aesthetic of the computer have software engineers been more inventive than in the proliferation of various kinds of screen-savers. Although this misleadingly pedestrian expression suggests things as useful and unglamorous as safety goggles and glare guards, screen-savers are a fascinating new form of electronic poetry whose intricate, fluctuating patterns have transformed the computer terminal into nothing less than a radical experiment in corporate art. Their basic style is that of Daliesque surrealism — floating clocks (with a choice of modern, antique, or digital, with, or without, a second hand) that actually tell time and that rebound off the monitor's invisible "walls," ricocheting back and forth in slow motion; or multisegmented worms in rainbow colors that slowly devour the screen in a ravenous feeding frenzy, munching away at the menus while the computer makes grotesque sounds of chewing and swallowing. As if defying the insipid silk screens, abstract collages, and ubiquitous textile wall hangings that contribute to the institutional atmosphere of the business world, software manufacturers have begun to subvert the staid conventionality of office interior decoration with a bizarre new type of animated screen art.

This recent attempt to aestheticize our terminals would seem to suggest that the generic and collective spaces in which we work are becoming much more permissive and idiosyncratic, tolerating new expressions of childishness, humor, and self-burlesque, images that undercut the stiff formality of the business world, its pompous gravity and dehumanizing preoccupation with appearances and protocol. In one of the most popular designs, a school of wingèd toasters fly across the screen in formation like a flock of migrating geese, flapping their exiguous wings as they dodge in and out of slices of floating toast, whose color the user can adjust according to his breakfast preference, from light to medium to dark. In another, brilliantly colored tropical fish, lobsters, sea horses, and jellyfish rummage around for food on the bottom of the ocean, scavenging through swaying thickets of seaweed that gurgle with underwater sounds emanating from the terminal. In yet another, swans glide on the surface of the screen, gracefully dipping their necks into the water to pluck out aquatic creatures, while in another the pitch-black monitor slowly begins to light up with the eyes of nocturnal animals, some slanted and feline, others round and surprised like an owl's, as if we were being observed by the ferocious inhabitants of a jungle, who stare out at us with unflinching curiosity as bloodthirsty wolves and coyotes howl in the distance.

Although these images would appear to be nothing more than the cinematic doodles of bored computer engineers, it would be a mistake to dismiss them as just the disembodied equivalent of the floral pattern on the stem of a spoon. A careful examination of the superfluous details that software designers build into their products reveals that the aesthetic of the computer reflects the psychology of the modern office. The freakish and visionary lyricism of screen savers articulates the state of mind of a work force experiencing unsettling professional insecurities which affect not only those who use the machines but those who create them. The surrealistic effects of software imagery are in part the byproduct of an occupational identity crisis occurring among the upper echelons of the labor pool, the computer intelligentsia, whose members are making a conscious effort to rid themselves of the denigrating stereotype of the inept, humorless misfit, the bumbling, ham-handed introvert. By filling our screens with zany non sequiturs, like droplets of jet-black "rain" that splatter over our work as if they had been flung at our terminals by Jackson Pollock, this rising class of educated specialists are attempting to prove to themselves, as well as to others, that, far from being nerdish automatons, they are in fact mad scientists and creative geniuses. The surrealistic humor of the screen-saver (the toasters swim as well as fly, dog-paddling through swarms of deep-sea creatures) is one of the secret handshakes of the new brethren of high technology, the cliquish fraternal order of those who are "in the know," who "get it."

It is thus representative of a new sensibility afflicting the electronics industry, that of ludicrous whimsicality, the kookiness of the oddball player of esoteric pranks. Mischievous software jocks now use the quirky and decidedly off-beat humor of new forms of computer imagery to shore up their own sense of group identity as an elite set of Silicon Valley absurdists. In one of the latest screen-savers, the monitor becomes a blackboard on which a piece of chalk begins to write two types of messages, either complex physical formulae, if the user chooses to be "Einstein," or repetitive lists of punitive precepts, if he chooses to be an "underachiever," with stern injunctions like "setting pigtails on fire is not performance art," "I will not xerox my butt," or "I will not barf unless I am sick."

Such compulsive wackiness serves two functions: it proves to the technician that he is a crazy nonconformist rather than merely an office drone, and, more importantly, it engenders respect among the lower echelons of the work force, who are fascinated by the conjuring tricks of a group of specialists whom they perceive as nothing less than a new breed of white-collar shamans. As insignificant as it seems, the "far-out" nature of the decorative features with which software designers are now adorning the computer provides one crucial piece of evidence of the growing stratification of the labor pool into two polarized factions in a rapidly evolving new caste system.

Unskilled white-collar workers are affected even more seriously by professional anxieties endemic to a society increasingly dependent on high technology. The startling contrast between the chaste elegance of the workplace and the gaudiness of screen-savers is symptomatic of a general trend in the computer industry to revamp antiquated corporate imagery in order to establish in the user's mind a healthy, if entirely spurious, connection between the irreconcilable spheres of business and fantasy, labor and recreation. By wallpapering our terminals with frivolous, technicolor images, reminiscent of Saturday-morning cartoons, manufacturers are undercutting our resistance to high technology and thus assisting in the assimilation of the computer into a culture so baffled by complex software programs that its fears about using them must be constantly placated with metaphors that exaggerate their accessibility.

The aesthetic of the computer thus masquerades as a form of decoration but is actually very useful. It pretends to entertain but in fact it teaches and, in many ways, it indoctrinates. Because the dismay caused by high technology has a potentially demoralizing impact on our work and could therefore cause inefficiency or even inspire resistance in old-fashioned employees frightened by the unfamiliarity of these newfangled gadgets, manufacturers actively infantilize the computer by means of a manipulative sort of anthropomorphizing rhetoric known by the duplicitous term of "friendliness." This revolutionary iconography is part of an industry-wide attempt to reconfigure the computer in accordance with the paradigm of the toy or the video game, thus coaxing the worker with puerile bribes and patronizing pats on the back that make him believe that his utensils are so elementary, so self-evident, that even a child could use them. Images of games form one of the basic themes of screen savers. In "Hall of Mirrors," the surface of the screen splinters into looking glasses that transform the terminal into a kind of fun house in which cavernous halls of infinitely regressive images tunnel back into the monitor. In "Marbles," small spheres are shot through an obstacle course of metal pegs where they bounce around like the silver balls of a pinball machine, and in "Puzzle," the screen becomes a grid of square pieces that are continually rearranged until the image on the monitor is slowly scrambled into a cubist collage of fragments of disjointed text and shards of broken mouse icons.

In the alienating world of office automation, the antidote to incomprehension is not education but a new kind of opium of the people, the ruse of "friendliness," whose gimmicks bring computers down to our level and encourage us to *forget* our ignorance rather than to overcome it. As the computer assumes an aura of childish intelligibility and performs for our benefit an act of self-degradation in which it talks baby talk rather than the technobabble of specialists, its "congeniality" drugs us into a soothing state of oblivion to the electronic riddles of a piece of sophisticated technology that we use with a mixture of uncomprehending pride and utter consternation. It is one of the major paradoxes of the business world that the more we lose control over our appliances, the more control we are deceived into believing we exercise over them, so that our ignorance of the abstruse secrets of transistors and circuit boards stands in an inverse relation to our growing sense of mastery.

The Pac Man aesthetic of the computer turns us, first into children, and then back again into adults — or at least into children who imitate adults by playing a game of dress-up in which we live out an occupational fantasy that our responsibilities, as seasoned technicians, the foot soldiers of the information age, are far less menial than in fact they are. By providing a multiplicity of choices for nearly every decorative feature offered, from the color of the monitor to the size of the design, manufacturers encourage us to manipulate the images on the screen and thus to achieve an illusory sense of being crack-shot telecommunications engineers. The malleability of the aesthetic of the computer transforms unskilled members of the work force into software virtuosos, artistic collaborators who, in the act of constructing these complex desktop dioramas, remake themselves in the image of their superiors: the hard-core techies — the "Systems Administrators" and the "Coordinators of Computer Services." Unlike the embellishments on toasters and colanders, which are by necessity inert designs, the decorative features of software are interactive. In the case of the screen-saver "Can of Worms," for example, the user can choose to make the worms' bodies either short or long, from between one to twenty segments, and, moreover, can make them move in a way that is either "wriggly," "straight," "crawly," "weavy," or a combination of all four. This bonanza of possibilities is surpassed only by "Mountains," in which you can display topographical maps of all nine planets, or "Rainstorm," in which you can adjust the screen's turbulent meteorology according to the velocity of the wind. As useful drills in the exercise of pure volition, computer decorations provide a form of mental calisthenics that bolster the self-esteem of a despondent underclass. They glamorize new high-tech jobs that, far from being the chance-of-a-lifetime opportunities that the electronics industry would have us believe it has created in great abundance in the service sector, are in fact just the usual numbing forms of drudgery trumped up to look more challenging and intellectual than they are by means of a psychological ploy basic to the aesthetic of the computer, the stimulant of false empowerment.

Computer imagery also invites us to view the machine as a compatriot, a brother-in-arms, a fellow cynic who shares the same jaundiced appraisal of the delicious aggravations of interoffice politics and who has the freedom to express out loud the boredom and frustration that we ourselves must always repress. Acting on our thwarted impulses to desecrate and disfigure our terminals, screen-savers are constantly turning the pristine, rational workspace of the monitor into the butt of an outrageous visual prank that elicits vindictive glee as we watch what appear to be deadly computer viruses attack and erase all of our data. In the screen-saver "Down the Drain," a hole develops in the middle of the monitor like the drain of a sink, an open sewer around which the entire screen begins to revolve like a waterspout, swirling dizzily into a vortex until it is sucked into an imaginary cesspool, as if someone had just pulled the plug or flushed the toilet. This impish spirit of destructiveness is also found in "Punchout," in which random shapes are carved out of the screen like puzzle pieces and spirited magically away, escorted offstage until everything disappears, stripped bare by unseen looters who plunder our monitors of every last scrap of text. The introduction of incoherence or even, in the case of the worms, decomposition into an arena that is always oppressively tidy and hygienic titillates the user by inciting within his terminal a miniature workers' riot in which he experiences the vicarious thrill of ransacking the screen, fouling his own nest, and sabotaging this inordinately expensive piece of equipment through an imaginary form of electronic terrorism. By inspiring a sense of community united by irreverence, by a shared bond of insolence and sarcasm, the aesthetic of the computer allows the worker to participate in a form of make-believe disobedience.

The decorative elements of software programs also subtly affect the worker's relation to his PC by providing frequent evidence of the human touch — the odd, superfluous detail that reassures the user that the computer was somehow "created" like a piece of art rather than "manufactured" like an appliance. These poetic "touches" constitute the equivalent in the electronics industry of the individualized workmanship that Ruskin and Morris admired so much in the carvings of Gothic cathedrals, whose intricate designs bore unmistakable traces of the personalities of their creators and thus represented a rejection of the mass-produced, assembly-line featurelessness of decadent modernity. A close inspection of the screen-saver "Marbles," for instance, reveals that when one of the five marbles represented, a sphere about 1/10-inch in diameter decorated with a Happy Face, caroms off the metal pegs and boomerangs back to the sides of the screen, its mouth opens up into an astonished expression of breathless surprise like a child squealing with delight on an amusement park ride. Details similar to this can be found in even the most commonplace software programs, like the sound of the game-show gong signaling a wrong answer that blares out like taunting laughter when the cursor reaches the very bottom of a document in Microsoft Word (called "clink-klank," it is one of several options that include sounds like "boing," "chime," and "monkey," the latter being a simulation of a shrieking chimpanzee). In order to accommodate the nostalgia and sentimentality of capitalism, which now continually harks back to the outmoded standards of production characteristic of the preindustrial era of the handmade and the homespun, an industry that constitutes the very summit of assembly-line standardization tries desperately to assuage the consumer's fears of anonymous machines by providing abundant evidence in their design of their inventor's whimsical presence. The aesthetic of the computer is thus rooted in the civil war that consumerism is now waging against itself.

At the same time that manufacturers mask the computer in three separate disguises — that of a toy, that of an irreverent compatriot, and that of a handmade piece of what might be called bionic folk art — they recognize the economic value of sweeping us off our feet with dazzling, sumptuous feasts for the eye that instill in us a sense of stunned incredulity. Even as they attempt to placate the worker and to prevent his alienation from affecting his job performance, they continue to intimidate him with images of their product's seemingly irrational powers, thus inspiring blind devotion, an attitude that serves an obvious function of strengthening the grip of high technology on the imagination of the American public. In order to perpetuate our admiration for the computer's miraculous capacities, they have developed an aesthetic that is entirely incompatible with the aesthetic of infantilization: the psychedelic. Hallucinatory images suggestive of the graphic style of the 1960s are fundamental to screen-savers that, for all of their futuristic pretensions, seem almost dated, like animated black-light posters from the era of acid trips, altered states of consciousness, and the peyote visions of a Carlos Castaneda. In one of the most popular software programs, intersecting loops of colored threads weave and unravel dense thatches of lines to form an image that exerts an hypnotic effect designed to elicit from the viewer such clichéd responses as "wow" and "far out," expressions that denote inarticulate amazement, the sort of speechless surprise we experience before natural wonders. Most screen-savers, like one of the all-time favorites, "LavaLamp," use the kaleido-scopic effects of a pop style specifically developed some thirty years ago to provide a visual analogue for mystical confrontations with the infi-nite — oscillating Möbius strips; stained-glass windows with ornate tracery that open up like buds blossoming in time-lapse photographs; protean shapes drawn with gyroscopes; mosaics that shatter and disin-tegrate; and uncanny microbe-like cells that divide and multiply in gaudy bursts of color — all of which link the computer with the per-ceptual distortions of mind-bending drugs.

Manufacturers also awe the consumer with literal images of the infinite: comets with iridescent tails that streak across intergalactic voids, photographic replications of spinning galaxies, or three-dimen-sional globes that hurtle through space rotating drunkenly on their axes. In "Warp!" the screen becomes the cockpit of the *Starship Enterprise* and explosions of supernovae rush by, as whole constella-tions are sucked into black holes or fan out, like the Big Bang, in expanding ripples of primordial matter. One of the central themes of the screen-saver is that the terminal is a trapdoor that leads into infini-ty, that the microcosm contains the macrocosm, that the monitor is identical to the sky, a Borgesian paradox based on the idea that this finite box of silicon chips and cathode-ray tubes in fact encompasses the entirety of the universe, easily housing all of the celestial bodies. The user responds viscerally to such dreamlike images, which induce in him a kind of giddiness, especially those that suggest that the terminal con-tains a parallel, "virtual" reality, a separate realm that lies perpetually out of reach, an illusion fed by irrational fantasies about omnipotent machines capable of performing instantaneously activities that would have taken him the better part of his day. The psychedelic appearance of new forms of computer decoration thus reflects the modern worker's constant state of silent amazement, which manufacturers cultivate as vigorously as they cultivate our snuggly sense of the machine's unthreatening familiarity, its "friendliness." With such economically advantageous and contradictory hyperboles, which simultaneously relax us and frighten us, beckon us closer and push us further away, the electronics industry enthralls and, in some sense, exploits the corporate environment.

# THE REPAINTING OF MODERN LIFE

*Marina McDougall*

# BANALITIES OF INFORMATION

It is modernity which has caused everyday life to degenerate
into "the everyday."
— Michel Trebitsch, Preface to Henri Lefebvre's
*Critique of Everyday Life*

Communication technologies have long been integrated into the
landscape of everyday life. While current advertisements depict
cellular phones and laptop computers in the business-gray hues associated with high-powered, transnational deal-making, these photos show
the ordinary reality of some contemporary mediated experiences.

Though spatial metaphors like "information superhighway" or
"cyberspace" attempt to place the ethereal interactions that transpire by
means of electronic appliances, the question remains whether a computer "desktop" is a desktop or whether the light emitted by a television
comprises a living world. One can be sure that familiar environments
— whether sidewalk, living room, or bowling alley — have been transformed (and often made more bleak and sensory deprived) by the
removed nature of telecommunications.

**"220 Volt"**
Beyond the controlled staging of department store displays or marketing campaigns, consumer electronics lose their glamor and become "pedestrian" on a hand-painted sidewalk sign. *p. 209*

**"You'll Like Me"**
The impersonal is given personality. *p. 210*

**Exercise Machine**
On a recent airplane trip an in-flight video promoted virtual-reality exercise machines. A treadmill is hooked up to a screen that depicts the sterile landscapes of computer-simulated parks. When the rider "turns" to the right or left or changes pedal speed, the screen jerkily flashes corresponding images that represent terrain overtaken. *p. 211*

**"You Can't Do This with Just any Camcorder"**
Depicting a police car on the camera's enlarged eyepiece, this billboard advertisement was made possible by the Rodney King tape — and its influence on events. The ad employs the lure of selling footage to "reality" television shows. *p. 212*

**Boy with Electronic Gun**
The player of this video game takes on the role of a rookie police officer. Animation depicts a car and suspect on the freeway from the point of view of a vehicle moving parallel. In a kind of drive-by shooting, the player accrues points when his shot hits the suspect. *p. 213*

**"(Don't) Do It, Powell"**
Closed-caption television displays the dialogue of a gang rape scene from the daytime soap opera, "One Life to Live." *p. 214-215*

**Virtual Reality**
Though players of virtual-reality games might have the feeling of being "somewhere else," onlookers see them as disoriented, batting about in empty space. This photo was taken at San Francisco's Cybermind — a virtual-reality gallery that combines elements of a hair salon and a gym. Bottled water was available, but no one seemed to be sweating. . . . *p. 216*

**Mr. Shy-D at ATM machine**
Like a street near a freeway on-ramp, the pavement near a bank of ATM machines has become a strategic spot to ask for a handout or to sell things. The ethereal, pseudo-private realm of electronic banking awkwardly intersects the living space of the homeless. *p. 217*

**The Real World**
Several dizzying layers of virtuality are depicted. MTV's pseudo–cinema verité series stages the interpersonal dynamics of a carefully selected group of twenty-year-olds living in a loft rigged with video cameras. One episode featured a trip to Rosalie's New Look, a San Francisco wig store that dates back to the 1950s; stills from "The Real World" are on display in Rosalie's windows. *p. 218*

**"Next Experience 13 Minutes"**
Sign outside a simulated earthquake attraction in San Francisco's Pier 39, a tourist area. *p. 219*

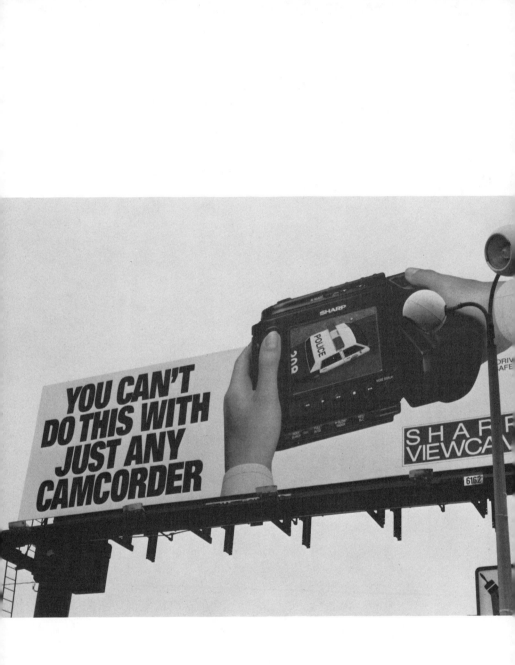

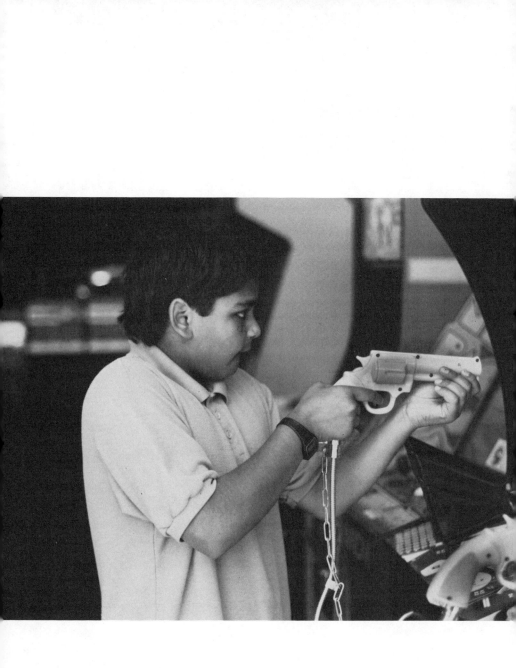

*Rebecca Solnit*

# THE GARDEN OF MERGING PATHS

You are in a maze of twisty little passages, all alike.
— Screen text from the early computer game, "Adventure" (Shallis 1984)

Place your right (or left) hand on the right (or left) wall of green, and
doggedly keep it there, in and out of dead ends, and you will finally get to
the middle.
— Julian Barnes on hedge mazes ("Letter from London,"
*The New Yorker,* September 30, 1991)

In 1989 I went to a demonstration at United Technologies in San Jose,
which was making fuel components for the nuclear-warhead-carry-
ing Trident II missiles. The corporate headquarters were nothing special, just another glass-walled box with Pizza Hut-style mansard roof, a
parking lot full of late-model cars, and nobody in sight but security
guards. It was in a business subdivision so new that much of the earth
was still exposed, raw compacted clay and gravel up to the curving suburban sidewalks, and there was a fruit orchard just behind the offices,
where one of the protesters escaped when chased by a guard. This, the
visible landscape of military technology, was bland, closed off, a mask.
There were other United Technology landscapes. Some were even more
invisible, or only potential: the military bases where the Trident missiles

were stationed, the targets they were intended for in this the late rococo phase of the Cold War, and the workplaces where they were manufactured — we were at design and corporate headquarters. (Nuclear
weapons are traditionally pork-barreled all over the country, so that
almost every state has an economic interest in their perpetration and no
one is responsible for *making* weapons.) Another United Technologies
landscape was underground, that of the colossal fuel plume that was
(and is) leaking toward the reservoir that held most of San Jose's drinking water. Although Silicon Valley's industries are often thought of as
clean because they lack industrial-era smokestacks and other such visible emblems of poisons, they are full of such high-tech toxins in the
workplace and in storage tanks leaching underground into the water
table. The most visible UT landscape at the time of our protest was an
ostentatious show of American painting, mostly landscapes, from the
Manoogian Collection in Detroit, underwritten by this corporation that
was destroying so many landscapes out of sight. The works in this show
at San Francisco's M. H. de Young Memorial Museum ranged from the
Hudson River School of the 1830s to American Impressionism at the
turn of the century, mostly heroic and idyllic landscapes, images of glorious possibility and pleasant interlude. This was what UT chose as its
public face.

Finding the landscape of Silicon Valley isn't as easy as getting lost among
the subdivisions and freeway exits and industrial parks. . . . When
Langdon Winner (1993) wrote a profile of Silicon Valley a few years ago,
he reached for the Winchester Mystery House as its emblem. It's certainly an obvious one in a region whose other landmarks are scarce. The
Stanford Linear Accelerator cosponsored by the Atomic Energy Commission, Paramount's Great America amusement park (with its Top Gun
military flight-simulator ride), Moffet Air Field, the off-limits Blue Cube
missile-control center next to Lockheed (officially called Onizuka Air
Force Base after one of the *Challenger*'s victims), Mission Santa Clara
— all contain something of the valley's character as well; but Mrs.
Winchester's paranoiac maze in San Jose sums it up best. Sarah Winchester
moved west after she became widow of the man whose repeating rifle
was the definitive weapon in western expansion — "the gun that won the
West." Frightened of the souls of the natives killed by the Winchester

repeating rifle, she sought spiritual advice and was told that as long as her house was being built she was safe — and the result is the 160-room chaos of architecture that has been a local tourist attraction since 1922. The house had no overall plan, so that doors and staircases lead nowhere, windows open onto rooms added later, architectural details clash, floor levels and design scales are inconsistent, and the workers went at it twenty-four hours a day so that it was always in process. Perhaps the house can be seen as a mad monument to mechanized capitalism. In the words of *Capital* itself: "If machinery be the most powerful means for increasing the productiveness of labour — i.e. for shortening the working time required in the production of a commodity, it becomes in the hands of capital the most powerful means for lengthening the working-day beyond all bounds set by human nature. It creates on the one hand, new conditions by which capital is enabled to give free scope to this its constant tendency and on the other hand, new motives with which to whet capital's appetite for the labour of others" (Marx 1932).

The invisible counterweight to the elaborate uselessness of this monument to wealth and fear is the ruthless efficiency of the rifle that paid for it: between the two of them — military technologies and diversionary follies — the valley might begin to be defined. The rifle's pursuit of death in open, contestable space; the house's sequestering from death and the dead in sequestered interior space. The implications of Mrs. Winchester's acts are interesting: that guns do kill people, that technology does have a moral dimension, and that perhaps she could buy her way out of the implications, fend off the spirit world with unending consumption, build a literal nowhere in which she could become lost to the spirit world.

What other stories can provide a thread through the labyrinths of Silicon Valley? The problem of understanding it seems to be in the inadequacy of its stories and images. There's the arcadian one of paradise lately become limbo, of the world's greatest prune orchard paved over to become the world's greatest technology center; and there's the utopian one of the glorious future being opened up by technology, the old Crystal Palace–World's Fair rhetoric that has become less credible for most people about most technologies. The two stories have some interesting things in common. The arcadian nostalgia of Wendell Berry or

Jerry Mander has its counterpart in the feckless utopian enthusiasm of the *Wired* and *Mondo 2000* consumers for a brave new world of cyber-space and techno-wonders. Mander's *In the Absence of the Sacred* (1991) is among the most recent attempts to assess technological progress, but the book bogs down in the author's refusal to engage social issues (as well as a romanticization of his own early years, in which the Great Depression becomes Edenic). Technology becomes an inevitable march toward consolidation, control, ecocide, a kind of Big Brother Godzilla. By making technology autonomous, rather than literally and historically a tool of power, Mander avoids most questions about the social forces that control the development and use of machines and the social changes that might detour us from their current trajectory. What begins as a radical critique ends as a refusal to engage the powers that be. In this, Mander is not much different than the more widespread enthusiasts for the new technologies, who also imagine technology as autonomous, and also leave out any social analysis, except for happy projections of empowerment through information access. Both these arcadian and utopian analyses insist on a straight line, backward or forward toward the good; but in a maze straight is the quickest route to immobility, and the road in may call for lateral moves, shifting perspectives. . . .

The maze becomes an inevitable metaphor for the moral tangles of technologies and social change, for the equivocal gains and losses, for arguments that can only lead deeper in, not outside the problem, for the impossibility of plunging straight forward or backing out altogether — that is, for simply embracing or rejecting the technologies and the visions of futures that accompany them. And the maze's image is echoed in the circuit boards and silicon chips, in the suburban sprawls of curving residential streets and industrial parks, of centerless towns that melt into each other, in the limited choices of computer games, perhaps in the rhetoric of technological progress that avoids social and teleological questions. Silicon Valley itself is an excellent check on the technophile's enthusiasm, since the joyous liberation of the new technologies is so hard to find here, in a place known for its marathon work schedules, gridlock traffic, Superfund sites (twenty-nine, the greatest concentration of them in the nation), divorce rate, drug consumption, episodes of violence, and lack of corporate philanthropy and organized labor (Winner 1993; Hayes 1989).

Certainly the orderly grid of fruit trees is more appealing than the jumble of mismatched corporations and assembly sheds, and certainly the most familiar story about California, even about America, is of a paradise that fell sometime not long ago, the story Mander tells. But the paradise of the orchards is partial at best: they are themselves work-places for immigrant and migrant labor whose poor working condi-tions and exposure to pesticides foreshadowed the sweat shops of microchip manufacture. And the first of these fruit trees came with the Spanish missionaries in 1777, who established Mission Santa Clara as a slave-labor camp for the Ohlone and nearby indigenous people. (Santa Clara County is named after the mission and includes San Jose and the southern half of Silicon Valley; the northern half extends up along the San Francisco peninsula into San Mateo County; and the term "valley" is something of a misnomer for this sprawl.) When the missionaries came on their double mission of salvation and empire, the whole penin-sula was a vast expanse of live oaks maintained by the Ohlone — the explorer Sir George Vancouver wrote after a visit in 1792,

> For almost twenty miles it could be compared to a park which had originally been planted with the true old English oak, the underwood . . . had the appearance of having been cleared away and had left the stately lords of the forest in complete possession of the soil, which was covered with luxuriant herbage and beau-tifully diversified with pleasing eminences and valleys. . . .
> (Jacobson 1984, 20–21; see also Margolin 1978; Hurtado 1988)

The orchards represent a reduction of a complex ecology into the monocultural grid of modern agriculture, and the transformation of a complex symbiosis with the land into the simpler piecework of agricul-tural labor for surplus and export. It may be that they even have some-thing in common with the Winchester repeating rifle as symbols of frontiers of conquest and rules of order. But they also represent suste-nance and continuity, two things hard to condemn out of hand, and I have been told that the sight of the valley in bloom was exquisite.

By the 1820s, the slave population — which included members of tribes from farther away as well as locals — had begun to escape, raid their former prison, and liberate their comrades. One successful raider, Yoscolo, carried out many such missions until he was caught and his head was nailed to a post near the church as a disincentive to the remaining workers (Jacobson 1984, 26). This is the not very edifying early history of European civilization in Silicon Valley, and the anti-colonial raiders here have their successors in contemporary Vietnamese gangs that steal vast quantities of silicon chips for the gray and black markets. Perhaps the missions, too, are prototypes of Silicon Valley, of information colonization. The neophytes, as the mission captives were called, were required to memorize and recite long lists of saints, prayers, and so forth that they were unlikely to have understood; salvation was a matter of having the right information.

In between the missions and the corporations, a golden age is hard to find and a fall is hard to postulate. Leland Stanford, one of the Big Four railroad barons whose government-subsidized rail monopoly made him a millionaire many times over, founded Stanford University in 1885 as a memorial to his dead son. The photographer Eadweard Muybridge invented high-speed stop-action photography here in 1873 to settle Stanford's bet that all a horse's feet were off the ground simultaneously at some point during a gallop; his invention is often considered the crucial precursor of motion pictures. Around that time the Bing cherry was bred here by Seth Lewelling, who named it after his Chinese cook — according to legend, in lieu of back wages. (It's worth remembering that the Silicon Valley region is now also a capital of genetic engineering, with giant Genentech and, again, Stanford University deeply involved.) Technological innovations continued in the region, including Philo T. Farnsworth's invention of the iconoscope tube, a crucial TV component, in the 1920s when the valley had nearly 125,000 acres in orchards; Charles Litton's San Carlos labs, which did war work, laser research, and more; and the refinement of tape-recording technology for Ampex and ABC soon after World War II. Moffet Air Field opened up in the 1930s and was for sixty years an important aviation research center. Ted Smith (1994b), the executive director of the Silicon Valley Toxics Coalition in San Jose, calls the place the greatest concentration of military-industrial sites in the country. Later, Stanford University became an ally of the electronics industry in much the way that nearby UC Berkeley took on nuclear-weapons research and lab management; Stanford Research Park was built on university land in the early 1950s as Stanford Industrial Park. Stanford electronics engineering students William Hewlett and David Packard invented the audio oscillator in 1938 and sold their first ones to Walt Disney for *Fantasia*. Long before Robert Noyce invented the integrated circuit — the silicon chip that gave the valley its name — military and entertainment technologies were already aligned on parallel paths.

In 1958 the Santa Clara planning department published a report that jumbled its metaphors interestingly: "Santa Clara County is fighting a holding action in the cause of agricultural land reserves. We are a wagon train, besieged by the whooping Indians of urbanization, and waiting prayerfully for the US Cavalry" (Jacobson 1984, 230). The cavalry had already arrived in the form of defense contracts that supported much of the research and development in the field, a connection that doesn't fit in with the image of the independent inventor or with the images of the planning department. The fruit orchards of Santa Clara, like the citrus groves of Orange County and the San Fernando Valley, are vestiges of a cleaner environment and lower property values. In a place such as Cupertino, with land prices up to a million dollars an acre, hanging onto farm land is difficult (though some farmers became wealthy enough by selling some of their land to cultivate the rest of it for pleasure). By the 1980s more than four-fifths of the agricultural land had become industrial or suburban space and only 8,000 acres of orchard stood, much of it between office buildings and clearly doomed. The Peninsula and San Jose were developed with little more foresight than Mrs. Winchester's house.

In this, Silicon Valley is not unique but typical of contemporary America. It is a decentralized diffused region: postindustrial, postcommunal, postrural, and posturban — postplace, but for the undeveloped western slopes and the undevelopable bay. As Langdon Winner (1993, 59) writes,

> Perhaps the most significant, enduring accomplishment of Silicon Valley is to have transcended itself, and fostered the creation of an ethereal reality, which exercises increasing influence over embodied, spatially bound varieties of social life. Here decisions are made and actions taken in ways that eliminate the need for physical presence in any particular place. Knowing where a

person, building, neighborhood, town, or city is located no longer provides a reliable guide to understanding human relationships and institutions.

As much as specific products — for the military, for business, and for entertainment, whatever that is — Silicon Valley seems to have generated prototypes of a more pervasive American future, one of dislocation. It has no center; rather than a city radiating bedroom communities that generate a coherent commute, it consists of myriad clusters of industry and housing, with commuters jamming in all directions at the beginning and end of every workday. As we discovered at protests there, Silicon Valley lacks centers that can function as social or political arenas.

I went to another demonstration at Lockheed Missiles and Space Corporation, the region's biggest employer and the prime contractor for Trident missiles, where there were no sidewalks, no focal points, no public spaces. In some sense, protest and community had been designed out of the place, and the workspace too had been suburbanized. Many of the Silicon Valley corporations are based in "campuses": attractive, diffused pseudodemocratic spaces that belie the traditional corporate structure within most of them, a design exemplified by the not very parklike Xerox PARC (Palo Alto Research Center). Diffuseness seems to have become an irreversible condition, in which both the consciousness and the place for consolidating individuals, for community is virtually impossible. Suburbia represents an early triumph of such diffusion, and the new technologies often seem to further it. Suburbia is a landscape of privatized space, of the division of home from work, the scenes of production both industrial and agricultural (and now informational) from those of consumption, a sequestering that has progressed with the shift from the public space of shopping streets to the private space of shopping malls.

There is the decentralization of anarchist direct democracy, in which power is everywhere; and the decentralization of postmodern control in which power is transnational, virtual, in a gated community, not available at this time, in a holding company, incomprehensible, incognito — in a word, nowhere. Mrs. Winchester's house is also a maze whose center was nowhere, and here it is important to distinguish types of mazes as well. The original myth of a maze is of the one Dadaelus built at Crete to hide the monstrous result of Queen Pasiphae's union with a bull, the Minotaur. Later mazes, such as the mazes on the floors of many medieval churches, symbolically compress and reconstitute pilgrimage, and the maze functions not as a tangle in which to lose things but a mandala in which to find them (the artist Paul Windsor recently mocked this tradition with a giant sand painting at the San Francisco Art Commission Gallery, which merged the Tibetan or Hopi mandala with the microchip). These mazes often have only one route to the center. The maze at Crete and the Mystery House apparently have no center; as such they are types of the new landscape of the suburb, the multinational, the subcontracted and subdivided, the faces of nowhere, in which it is impossible to get found.

Here it is important to distinguish between the actual tools generated in Silicon Valley and its sister sites from the visions of their implementation. Computers and the information they manipulate are the means to many ends; in one of these they are an end in themselves. In its most dematerialized state Silicon Valley is a blueprint for a future: in this future outside has disappeared, the maze has no exit. The world of information and communication on line, much hailed as a technological advance, is also a social retreat accompanying a loss of the public and social space of the cities, the aesthetic, sensual, and nonhuman space of the country, a privatization of physical space and a disembodiment of daily life. A central appeal cited for the new technologies is that their users will no longer have to leave home, and paeans accumulate to the convenience of being able to access libraries and entertainments via personal computers that become less tools of engenderment than channels of consumption. This vision of disembodied anchorites connected to the world only by information and entertainment mediated by the entities that control its flow seems more nightmarish than idyllic. Postulated as a solution to gridlock, crime on the streets, the chronic sense of time's scarcity, it seems instead a means to avoid addressing such problems, a form of acquiescence.

There is another maze, another landscape, which has bearing on the tangle of Silicon Valley, the multimedia mazes that resemble the maze of Jorge Luis Borges's "The Garden of Forking Paths" (1970), in which a Chinese assassin finds out the secret of his ancestor's chaotic novel and missing maze — the two are one:

Ts'ui Pên must have said once: *I am withdrawing to write a book.* And another time: *I am withdrawing to construct a labyrinth.* Everyone imagined two works; to no one did it occur that the book and the maze were one and the same thing. . . . Almost instantly, I understood: "the garden of forking paths" was the chaotic novel; the phrase "the various futures (not to all)" suggested to me the forking in time, not in space. . . . In the work of Ts'ui Pên, all possible outcomes occur; each one is the point of departure for other forkings.

An extensive but finite number of forks can be represented on an interactive CD or laser disk, but they do not reproduce life, in which the unimaginable is often what comes next; and the greatest tragedy of the new technologies may be their elimination of the incalculable — of the coincidences and provocations and metaphors that in some literal sense "take us out of ourselves" and put us in relation to other things. To live inside a mechanical world is to live inside plotted possibility, what has already been imagined; and so the technologies that are supposed to open up the future instead narrow it. I am not arguing for existentialist freedom with this difference between inside and outside, only for an unquantifiable number of paths in the latter, a too predictable course in the former.

Much recent attention to interactive media proposes that it makes the passive viewer become actively engaged. What is interesting about these products is that they map out a number of choices, but the choices are all preselected (and with the rare exception of such artists' work as Lynn Hershman's, the choices have little to do with meaningful decisions). That is, the user cannot do anything or go anywhere that the creator has not planned; as usual with computer programs, one must stay on the path and off the grass (by which analogy hackers do get off the path, a subversive success that keeps them in the park). We could chart the game as a series of forks in the road, in which each choice sets up another array of choices, but the sum total of choices has already been made. Thus the audience becomes the user, a figure who resembles a rat in a conceptual version of a laboratory maze. The audience-user is not literally passive; he is engaged in making choices, but the choices do not necessarily represent freedom, nor does his activity represent thinking.

Participating is reduced to consuming. The ur-game, Pac Man, made this all apparent: the sole purpose of the Pac Man icon, a disembodied head-mouth, was to devour what is in its path as it proceeded through a visible maze.

Perhaps what is most interesting about this form of interactivity is its resemblance to so many existing corridors of American life, in which a great many choices can be made, but all choices are ultimately choices to consume, rather than to produce. About a decade ago, the 7-11 chain ran a series of television ads whose key phrase was, "Freedom of choice is what America is all about." The ads echoed a pervasive tendency in the culture to reduce freedom to the freedom to choose from a number of products, to the scope of the consumer to consume. Perhaps it is not surprising that consumption should become the metaphor for democracy in a country that has long had little but representative democracy: that is, the ballot too is a kind of garden of forking paths and not an open plain on which to roam and encounter. That is, by the time the political process has reached the voting booth, all the real choices have been programmed in, and the voter becomes a consumer. Few genuine choices remain, and the act of voting becomes the act of acquiescence, endorsement of the maze as an open field. The laboratory maze through which the rat moves is one metaphor for it. Another is supplied by the critic Norman M. Klein writing on virtual reality (VR): "VR is reverse Calvinism — predestination posing as free will. In that sense, VR may be as old as the Massachusetts Bay Colony, a new consumerist form of metaphysical redemption" ("Virtually Lost, Virtually Found: America Enters the Age of Electronic Substance Abuse," *Art Issues,* September-October 1991).

The real landscape of Silicon Valley seems wholly interior, not only in the metaphor of the maze and the terrain of offices and suburbs, but in the much-promoted ideal of the user never leaving a well-wired home and the goal of eliminating the world and reconstituting it as information. Again, what disappears here is the incalculable, this time as the world of the sensory and sensual, with all the surprises and dangers that accompany it. In all the hymns to information, little is said about the nature of that information or the ability to use it; one pictures the empty trucks of metaphor hurtling down that information highway.

Understanding works largely by means of metaphors and analogies — the incalculable relationships between bits of information — and the way those metaphors and analogies are drawn from the nonconstructed world. The most obvious examples are expressions: *stubborn as a mule, dumb as two sticks, pigheaded, dog breath, pussy, cock, cuckoo, horse sense, drones, worms, snakes in the grass, aping the gentry, bovine, donkey's years.* There are also shared (but fading) fables: the ant and the grasshopper, the tortoise and the hare, the dog in the manger, and a million coyote stories, which provide animal analogies for human dispositions, moralities, and fates. The microcosmic/macrocosmic metaphors are particularly important, and they're most immediately obvious in geography metaphors: the foot of a mountain, the bowels of the earth, a river's mouth, the heart of the forest, tree limbs, even the soft shoulders of roads. (For a minor example, in *Tristes Tropiques* Claude Levi-Strauss compares speaking of his researches to an unreceptive audience to dropping stones down a well, an analogy few would be likely to make nowadays.) The majority of figures of speech that make the abstract concrete and the abstruse imaginable are drawn from animals and organic spaces. It's the animal world that makes being human imaginable, and the spatial realm that makes activity and achievement describable — career plateaus, rough spots, marshy areas. And it's the image of the maze that's gotten me through all the aspects of Silicon Valley I've approached thus far, and the approach to a specific landscape in California that's made it possible to articulate some effects.

Finally, even nowhere has its twin: everywhere. Silicon Valley has become a nowhere in the terms I have tried to lay out — an obliteration of place, an ultimate suburb, a maze in which wars are designed, diversions are generated, the individual disembodied. But the physical landscape of Silicon Valley is now everywhere, not only in the attempts to clone its success, but in the spread of its products and its waste throughout the globe, the outside world being ravaged by the retreat to the interior. If you imagine a computer not as an autonomous object, but as a trail of processes and effects and residues that leave their traces across a global environmental maze, then it is already everywhere. The clean rooms in which poorly paid chip-makers were exposed to toxic chemicals are now subcontracted out in the Southwest, Oregon, and the Third

World, so there's a little of the valley there. The waste that was leaching through the once fecund earth of Silicon Valley is leaching still, and more of it is leaching around the globe. Some of the chemicals used to clean the chips further have been peculiarly potent ozone-depleters (though many Silicon Valley firms have switched over to other compounds, some of these substitutes are themselves harmful to the ozone layer), so think of the upper atmosphere too; and of the landfill where the packing and shipping material went; and the electrical generating station you're plugged into and its energy sources — coal, hydropower, nuclear, geothermal, natural gas? — think of the networks it may be hooked up on; think of the corporations whose pockets it lined — but don't picture pockets, the money is in imageless cyberspace — and the stock markets where their shares are traded; think of the forests the manuals are printed on; think of the store that sold it; think of where it'll be dumped when it's rendered obsolete, as all computers have been. These are the tentacles, the winding corridors, the farthest reaches of Silicon Valley, and the hardest to imagine. It is the scene of the crime that has vaporized, and resisting an unlocatable and unimaginable crime is difficult. One of the principal challenges for environmentalists is making devastation that is subtle and remote seem urgent to people with less vivid imaginations, another is finding a site to protest at (which is why Greenpeace has largely relocated from actual sites to wherever the media is). And the ultimate problem of the landscape of Silicon Valley in its most abstruse, penetrating, and symbolic forms is that it is unimaginable (see Hayes 1989; Smith 1994a).

Apple Computer, which is headquartered in six buildings indistinguishable but for their security levels on Infinity Loop in Cupertino, is a key landscape for Silicon Valley, one that apparently displaced real orchards. When I was there, the Olson orchard across Highway 280 in Sunnyvale was selling Bing and Queen Anne cherries, and Latino workers were cutting up apricots to dry; but a third of the orchard was bulldozed this spring for housing, and the rest of the Olson orchard is on its way out ("Last Call for the Last Sunnyvale Orchard," *San Francisco Chronicle*, August 1, 1994). What does it mean, this rainbow-colored apple with the bite taken out of it, which appears everywhere on Apple computers and on the many commodities (mugs, key-rings, T-shirts) Apple corporation markets, this emblem that seems to sum up the

Santa Clara Valley's change from agriculture to technology? It seems to have been appropriated to connote simplicity and wholesomeness, though apples aren't rainbow-colored in anything but the sloppiest association of positive emblems; and the bite also recalls temptation in Eden: the emblem is denatured, reassuring, and threatening all at once. But more than that, it is forgettable, dead in the imagination, part of nowhere — it has been a decade since I last pondered the Apple logo, which has become part of a landscape of dissociation in which the apple image connotes neither sustenance nor metaphor, only a consumer choice, the fruit of the tree of information at the center of the garden of merging paths.

*References*

Borges, Jorge Luis. 1970. "The Garden of Forking Paths" In *Labyrinths*. Harmondsworth: Penguin.

Hayes, Dennis. 1989. *Behind the Silicon Curtain*. Boston: South End Press.

Hurtado, Albert L. 1988. *Indian Survival on the California Frontier*. New Haven and London: Yale University Press.

Jacobson, Yvonne Olson. 1984. *Passing Farms, Enduring Values: California's Santa Clara Valley*. Los Altos: W. Kaufmann and California Historical Ctr. [Jacobson is the granddaughter of the founder of the Olson orchards.]

Mander, Jerry. 1991. *In the Absence of the Sacred: The Failure of Technology and the Survival of the Indian Nations*. San Francisco: Sierra Club.

Margolin, Malcolm. 1978. *The Ohlone Way: Indian Life in the San Francisco–Monterey Bay Area*. Berkeley: Heyday Books.

Marx, Karl. 1932. *Capital, The Communist Manifesto, and Other Writings*. New York: Modern Library.

Shallis, Michael. 1984. *The Silicon Idol: The Micro Revolution and Its Social Implications*. New York: Schocken Books.

Smith, Ted. 1994a. "Coming Clean in the Semiconductor Industry." Interview with Anita Amirrezvani. *Bay Area Computer Currents* (June 1–13).

———. 1994b. Telephone interview with the author (September 8).

Winner, Langdon. 1993. "Silicon Valley Mystery House." In *Variations on a Theme Park*. Edited by Michael Sorkin. New York: Noonday.

*Chris Carlsson*

# THE SHAPE OF TRUTH TO COME:
# New Media and Knowledge

E verything we know about entertainment and the forms it takes as
"product" is up for grabs. The categories that seem so "natural" to
us — TV, radio, albums, books, magazines, movies, and videos — are
rapidly converging into one large digital data stream. Those earlier
forms won't completely disappear, but all will be altered by their new
interchangeability as data, and new combinations will become common. Central to this process are converging changes in form and delivery, from the much-touted arrival of "interactive" media to the frenzy of
corporate and legal deal-making regarding the delivery of digital signals
to your home or business via phone or cable.

Beneath the media world lies our perceptual framework, and digital media may change how we *know* what we know. Our sense of life
and society changed at earlier times of upheaval in "communications
technology," especially in the transition from oral to literate cultures.
Literacy contributed to the downfall of many a dictator and monarch,
but it also brought with it certain assumptions that strongly influence
our imaginations. Marshall McLuhan argued that the subtle effects of
the medium of knowing influences what we can know. Knowledge,
when constructed from "straight rows of exactly repeatable, individual-

ly meaningless units of type, is an amazingly close analogue of, and perhaps the model for, the specialized industrial society in which an entire economy is assembled out of small bits of individually owned private property — including intellectual property" (Brent).

Any metaphor can be taken too literally, but clearly something as invisibly "natural" as the alphabet imparts deep assumptions about how the world around us is structured, or more accurately, how we humans structure that world. Literacy provided the "operating system" and the logic for the advanced developments in communications technology by establishing the basis for a technologized culture and by shaping our conception of knowledge.

The subversive possibilities of literacy per se have long ago exhausted themselves. Seeing the world through literate eyes, as does a large part of the world's population, has not in itself led to a richly engaged and informed public, even though books and information are relatively easy to acquire. The critical consciousness of an active literate (still pretty rare, after all) has been outflanked by the aggressive shaping of "reality" by mass media. Of course there could be no TV without literacy, but the represented world of television, reinforced by radio and newspapers, establishes and shapes reality in ways that the printed word only aspired to, but could never achieve alone. After centuries of gradually expanding literacy and nearly a hundred years of public schooling, our minds have been shaped to believe what we see. As photography, film, and TV became commonplace, our "natural" instinct to believe what we see created a society perfectly suited to "blind" allegiance to a carefully manufactured "reality" of images. The roots of this manipulability are clearly visible in the successes of yellow (print) journalism around the turn of the century before the arrival of the "more real" radio or TV.

If the demise of the Soviet empire heralds the end of the twentieth century, it also marks the victory of the system of order advanced by the U.S. throughout the world, a system Guy Debord (1990) called the "integrated spectacle":

> The society whose modernization has reached the stage of the integrated spectacle is characterized by the combined effect of five principal features: incessant technological renewal, integra-

tion of state and economy, generalized secrecy, unanswerable lies, and eternal present. . . . Once one controls the mechanism which operates the only form of social verification to be fully and universally recognized, one can say what one likes. The spectacle proves its arguments simply by going round in circles: by coming back to the start, by repetition, by constant reaffirmation in the only space left where anything can be publicly affirmed, and believed, precisely because that is the only thing to which everyone is witness. Spectacular power can similarly deny whatever it likes, once, or three times over, and change the subject, knowing full well there is no danger of any riposte, in its own space or any other.

The rule of the spectacle, while omnipresent and amazingly effective, still has its cracks and fissures. Clearly, the simple truth no longer holds the same weight as it once did, and it never seems "simple." Reliance on earnest appeals to the truth will continue to fall on deaf ears, if such appeals are even articulated publicly at all. The "new media universe," or media-verse, is under construction, and some people hope for openings in the armor that will allow a more egalitarian society to begin emerging from the technological cornucopia.

## Predatory Pruning in the Corporate Garden

Corporate giants have recently been observed tying the knot in frenzied cross-industry deals, getting married to stake a claim in the much-anticipated media-verse. The old TV networks, Microsoft, IBM and Apple, TCI and Time Warner, the *New York Times* and *USA Today,* Bell Atlantic and the other Baby Bells, AT&T, QVC and the Home Shopping Network, not to mention the smaller local interests, have joined the battle. Vast fortunes will be wasted and a few will survive and grow. And when the dust clears there should be, according to the analysts, a media industry straddling the globe comparable to the mid-twentieth-century auto and oil giants. (The current maneuvering is reminiscent of the corporate conspiracy in the 1940s and 1950s to scuttle intracity urban rail systems when a cabal of General Motors, Firestone Tires, Phillips Petroleum, Mack Truck, and Standard Oil of Ohio bought up rail systems and "modernized" city transit systems by ripping out tracks and replacing trains with buses. The goal was to get people off public transit and into private cars, a plan that worked rather well. But much more is now at stake. The manufacture and maintenance of images of global reality may be even more powerful than the establishment and control of a highly profitable, carefully controlled, enormously wasteful, and finally doomed transit racket.)

As media giants compete across the planet to control our perceptions of reality, the univocal, self-referential spectacular society will have to change its spots. While we watch and throw an occasional stone, the system will try to exploit regional differences even while promoting a less Euro- or Yankee-centric "objectivity." The contest of CNN against ABC against BBC against TV GLOBO against NHK will supposedly demonstrate the "freedom of the airwaves." Competition will be emphasized to obscure the essential sameness and increasingly homogenized package of modern life, a package that is paradoxically very different from the lives of most people.

We can expect the approaching international network television system to promote a new global citizenship. How shall we counter this bogus citizenship, this pathetic acquiescence to a corporate agenda? What would an anticapitalist, positive, and humane version of such "citizenship" consist of in the postmodern world? Can "global citizen" or "international proletarian" or any new global identity arise to undermine the untrammeled power of multinational capital? Multinational corporations will spend billions to define a "desirable" way of life, ideologically reinforcing "globalism" the same way national capital has historically reinforced nationalism. Global broadcasting will surely intensify the already advanced process of creeping monoculture, leading to the final airport-ization and enclave-ization of reality for the haves, while the have-nots remain unseen and unnoticed, except as panhandlers, rioters, and tribalists.

The media owners will try to get us to pay for this new media-verse, too. Unless we can revolutionize how we use these technologies — along with the society we create together — they'll invent yet another payment scheme: by the minute, by the product, by the kilobyte, subscriptions and access fees, TV-shopping taxes, and so on. We can't play unless we pay, as usual, unless the easy duplication of digital information inadvertently undermines all attempts at ownership and payment schemes.

It is possible that the private origin and rightful ownership of ideas will erode as we freely access bits of writings through new electronic libraries. Someday we may realize that the global reservoir of scientific and technical knowledge belongs to everyone equally, because it is a product of the complex web of human history. Doug Brent argues that

> The metaphorical meaning of print technology is isolation, not communality. In particular, the ability to claim one's particular share of the intertextual web, and stamp it with one's own name — an ability made possible by the same printing press that made widespread cumulation of knowledge possible as well — suggests that knowledge is individually owned. I believe that computer mediated communication provides a totally different metaphorical message . . . that takes theories of collaborative knowledge and . . . stamps them indelibly in the consciousness of the entire society. . . . With electronic communication the notion of the static and individually owned text dissolves back into the communally performed fluidity of the oral culture. . . . Document assembly becomes analogous to the oral poet boiler-plating stock phrases and epithets into familiar plots . . . it becomes obvious that *originality* lies not so much in the *individual* creation of elements as in the *performance* of the whole composition. [emphasis added]

## Orality and Literacy

> Oral, non-literate cultures are "verbomotor" cultures in which, by contrast with high-technology cultures, courses of action and attitudes toward issues depend significantly more on effective use of words, and thus on human interaction, and significantly less on non-verbal, often largely visual input from the "objective" world of things. . . . Primary orality fosters personality structures that in certain ways are more communal and externalized, and less introspective than those common among literates.
> — Walter J. Ong, *Orality and Literacy* (1982)

Try and imagine life without books, magazines, packaging, signs, TV, radio, boom boxes, and so on. Not an easy task. What was "in" the pre-literate mind? What did it make of time and space?

Before writing and before alphabets, human society depended entirely on speech and song to establish and maintain knowledge, often in the form of lengthy, elaborate sagas. Ong argues that "without print, knowledge must be stored not as a set of abstract ideas or isolated bits of information, but as a set of concepts embedded deeply in the language and culture of the people." Oral cultures strive to conserve knowledge, largely through the repetition of elaborate allegorical tales with stock phrases and communally recognized characters, roles, and concepts (Homer's *Iliad* and *Odyssey* are examples). With no place to "look it up," humans depended on wise individuals, often the elders who had developed and polished their storytelling skills, to maintain and transmit what was known.

Intellectual experimentation was distrusted, since the important goal was to preserve what was known rather than to challenge and undermine it with new ideas. The oral society maintained a heightened awareness and focus on what we call "the present." Changes in life were reflected in changing episodes in well-known stories. What was remembered gradually and seamlessly changed to meet new situations.

Accompanying the move from an oral to a literate human culture, sound is demoted as the primary medium for experiencing knowledge. Sound, whether voice or ambient noise or music, surrounds you in a way visual input doesn't. Learning, understanding, and wisdom were by definition socially developed and shared — thoughts and wisdom only existed as sound, disappearing as soon as uttered unless repeated.

The "natural" separation between knowledge and knower that permeates our literate, technologized culture, couldn't have occurred to an intelligent, "well-educated" mind in a preliterate society. Oral people didn't have the sense of time we accept today. There were seasons and weather, and most cultures had holidays during the course of the year, but there weren't dates, hours, clocks, and so on. Literacy made it possible to record thoughts and examine, debate, and revise them, which soon gave greater power to the masters of the newly technologized word. As thinking about something written becomes more common, fragmentation of consciousness, specialization, and complex analysis become possible in ways not possible in oral societies. Among the earliest uses of writing were the control of legal codes and accounting for business. As market relations inexorably spread through imperial conquest and subjugation, literacy went along, too. Literacy, based on visual linearity, after centuries has narrowed what we value as knowledge, and hence what we experience. Even though we have more and deeper knowledge about the world than preliterate, oral cultures, our civilization is astonishingly barbaric. The everyday communality and ability to live much more cooperatively, present in many oral societies, would be a welcome antidote to the isolation and anomie of modern daily life.

### Interactivity to the Rescue?

The hype about interactivity suggests that certain appetites or consumer demands are being felt. Do people want interactive entertainment because the interaction they share with friends, family, and coworkers is insufficient? Now, all-new interactive entertainment comes along to assuage the loneliness of modern life, but ends up reinforcing it. Capitalist society brutalizes us with the fears and doubts of "economic necessity." We react *naturally*, by becoming more machine-like. Interactivity promises to return us what we lost as we adapted to society. We may be bored and boring, but interactive entertainment promises to let us control — in a glorified system of multiple choice — beautiful people doing beautiful things, with no back talk or guff.

Interactive programming will have to be able to deliver specific consumer market segments to advertisers, of course. Interactivity and artificial environments ("virtual reality") will attract a share of the entertainment consumer dollar. How much depends on what the experience can really deliver. If it ends up being a wax-museum trip through Polygon Hell, it will never catch on. But what if you could "attend" historic moments, places, events, and "be there" in true 360-degree live animation? (Advertisers will no doubt anachronistically slip modern products into the historic moments for added impression, as well as added revenue for the programmers.)

Some boosters argue that interactive programs can stimulate a renaissance in education, overcoming the archaic forms of learning still relied on in most schools. A great deal of public school is really awful, so it's easy to imagine a new series of techno-fixes being well received by students and faculty. But the issues of education go a lot deeper. For example, no technology can address the mysterious relationship between education and the economy — that is, *what do we really want our children to learn?* How does this educational system reflect the real priorities of this society, as opposed to its rhetorical claims, and will interactive educational tools reinforce those priorities or undermine them? But that's a subject for another essay. . . .

True interactivity is what happens between human beings, genuine subjects, individuals with the unique quality of being able to find a nearly infinite range of responses to any situation, as well as the ability to imagine completely new, unanticipated possibilities. Any interactive program or game today is a closed loop in which all the possibilities have been thought of and planned for; your "job" is to try to gain access to them. With a "friendly" interface, your work seems like play, and the time spent computing seems really fun and just a big game after all. But the interaction is the means to personalize and enhance your participation in prefabricated image consumption. By providing limited choices, interactivity mimics shopping and the false control offered over work by self-management and workers' participation schemes, wherein workers decide how to accomplish the business's mission, but, crucially, *not what the mission is.*

The free communication spaces that we have now (e.g., Internet, public-access TV) are already boring — because community is weak. The whole notion of public opinion has turned into an easily manipulated series of statistical non sequiturs. In Debord's words, "Unanswerable lies have succeeded in eliminating public opinion, which first lost the ability to make itself heard and then very quickly dissolved altogether."

The wide expansion of channels and bandwidth along with easy, cheap, two-way and conferencing capabilities *could* promote horizontal communication in ways that undercut the univocal voice of the dominant society. But the spectacle could also continue to absorb every social expression and movement into its underlying logic of buying and selling. The advent of TV shopping and on-line services expands the reach of market relations even further. Perhaps the loss of public space has driven the dreamers into cyberspace, with the only thriving "public" communities found on bulletin boards; hence the enthusiasm for new media in projects of social liberation, with little thought of the majority of the world's population that has simply been closed out of communication technologies, except as passive receivers.

Although e-mail and electronic discussion groups are bringing together new communities around shared ideas and interests, the people involved remain very isolated. The millions of Internet users are mostly very alone as they "communicate," so it's difficult to see how underground communities can develop to reclaim the public space essential to a free society. It's easier to imagine a lot of empty, pointless verbiage flying around the electronic world, matched only by the piles of data gathered by our corporate and governmental institutions.

### Reconnecting the Circuits

> In this world which is officially so respectful of economic necessities, no one ever knows the real cost of anything which is produced. In fact the major part of the real cost is never calculated; and the rest is kept secret.
>
> — Guy Debord (1990)

Dissenting views are virtually invisible in mainstream America. Broadcast television, malls, and airports comprise "public space" for most people, and have produced a life where "images chosen and constructed by someone else have everywhere become the individual's principal connection to the world he formerly observed for himself. . . . [It is] a concrete experience of permanent submission" (Debord 1990). In exchange for our self-doubt, the spectacle reassures us that we are sharing in "real life" when we watch it happen on TV. After all, the representation of life is "more real" than life itself.

Spectacular society leads us to dismiss our own experiences when it diverges too far from the official story. For example, the sustaining energy of the anti–Gulf War demonstrations in U.S. cities was in part drained by trivializing, limited media coverage. In San Francisco, 100,000 antiwar protesters were just another "opinion" alongside 300 prowar protesters in the suburbs. The reality of living through such a large demonstration became hard to believe when it was not reinforced in the "real" public sphere, TV.

In keeping a profligate consumer society based on increasingly sharp class divisions and falling living standards from im- or exploding, the world makers have a difficult task. They must allow a decentralization in spectacle maintenance. They have to assume that the principles of spectacular society (mistrust of one's own experience, suspicion of other people's motives, belief in the bald-faced lies of the rulers, loneliness, resignation, and atomization) are so thoroughly internalized that most people maintain these principles independently of any overt centralized control.

New media tools like "morphing" and photo-manipulation software have drastically eroded verifiability through images. (Morphing is a software process that transforms one face or figure into another.) The ability to manipulate consciousness and the appearance of reality has eroded with the loss of image believability. The development of interactivity is an attempt to outflank the increasing emptiness of media consumption by using *our participation* to enhance the credibility of spectacular images. More importantly, the new media seek to perpetuate the form of media commodity and suppress direct, horizontal, free communication.

Finally, this is what we face: to take the disparate strands of knowledge, culture, and meaning that we develop in our electronic activities — and elsewhere! — and give them a life in the physical and political world. We must remove the constraints of isolation imposed by our "interactive solitude." The threads of subversion we weave so quietly today must find their way to transform the self-destructive, brutal, and dehumanizing lives we lead at work, at school, and in the streets. The trust we place in electronic links must again find a common home among our social links, until electronic "experiences" take their rightful place as supplements to a rich, varied human life.

*References*
Brent, Doug. n.d. "Speculations on the History of Ownership." *Intertek* 3.4.
Debord, Guy. 1990. *Comments on the Society of the Spectacle.* Translated by Malcolm Imrie. London: Verso. [Originally published as *Commentaires sur la société du spectacle.* Paris: Editions Gérard Lebovici 1988.]
Ong, Walter J. 1982. *Orality and Literacy.* London: Routledge.

*Chris Riding*

# DROWNING BY MICROGALLERY

Great use of technology and interlinking screens; the trouble is you
spend so much time in here that you forget the actual gallery.
. . . .
Certainly better than yet another McDonald's.
— from the comments book for the MicroGallery
at the National Gallery, London

It only takes a glance through the comments book for the Micro-
Gallery — the digitized representation of the paintings in the British
National Gallery — to realize that this foray into multimedia has been
met with great approval from the public. As Ben Rubinstein, of
Cognitive Applications Ltd, the designers of the MicroGallery software,
stated in his presentation of the system to the International Conference
on Hypermedia and Interactivity in Museums (Pittsburg, October
1991), "Reactions to the system have been overwhelmingly favorable —
and frequently more enthusiastic than we ever expected." His presentation goes on to quote eight entries from the first few pages of the visitors' comments book that testify to the system's enhancement of the
museum-goer's time.

On the face of it, there is little more to be said: the system enhances the museum-goer's walk. It is certainly better than another McDonald's and, yes, the time spent in front of the computer screen could induce you to forget the actual gallery. The MicroGallery is absorptive in ways that the process of walking, stopping, and looking at the paintings themselves may no longer be. This is because the system offers you the lure of the labyrinth: it is like playing Tetris with image and text. Welcome to the salon of Super Mario.

Bernard Sharratt, writing in the *New York Times Book Review* (March 6, 1994) on the newly released CD-ROM version of the gallery ("Microsoft Art Gallery"), makes the analogy between the system and the arcade game:

> The response of the youngsters at the gallery suggests [treating] the CD-ROM as akin to an arcade game with the numerous exiting levels to explore. And the only alien to zap is art itself, or rather that restrained "high seriousness" still so firmly characteristic of the conscientiously gallery-trained.

In one sense this is true: it's possible that the paintings themselves are, in Sharratt's words, being zapped. Although time spent shifting through the on-screen allegory allows one to browse in art's arcade, the logic and impulse in moving through the publication belie the zoned worlds of most arcade or computer games. Here, information is intended as the key to mediate the "world" of the actual painting and its historical representation and fiction: in other words, the ultimate level for the viewer to explore should be the material makeover of the picture or the illusion of the pigment itself. Needless to say, it isn't that simple, as a consumer testimonial from July 13, 1991, bears witness: "Quite extraordinary. I could play here for hours. If you're not that careful it could turn out to be better than the actual paintings." This may be the case. . . .

The MicroGallery was opened as part of the new Sainsbury Wing on July 9, 1991. The facts are as follows. It is located next to the restaurant and below the Gallery's early Renaissance collection. The system has an 80 percent occupancy rate at all times with, on average, 200 people per day passing through the MicroGallery to use one of the twelve terminals. Each terminal has a nineteen-inch touch-sensitive screen, offset at an angle into the right-hand wall as you walk in. Nine of the terminals have a built-in printer, thus enabling the visitors to print out biographical, historical, and geographical information or plan their own individual iconographic programs. Once the information has been located on the computer screens of the MicroGallery, it can be printed out for a small fee and this extracted hard copy taken by visitors on their brisk walk through the National Gallery's cultural holdings. The printing charge is comparable to normal photocopying prices, but is in black and white only. During busy periods, each visitor is allowed twenty minutes at a machine, though there is an unpublicized booking system through which one can prearrange an extended time slot by calling MicroGallery in advance of one's visit.

The MicroGallery (and the CD-ROM) contains in the region of 4,500 pages: over 300,000 words of text and about 12,000 illustrations. There are also a number of animations related either to theme or individual painting. The average time for moving from page to page is around a second. The MicroGallery project took two and a half years to complete, and its sole sponsor is American Express, which contributed $1,000,000 to bring the system from inception to completion.

There are two drawbacks to the MicroGallery, and these are echoed in the visitor's comments book: first, the lack of color printing and, second, the accessibility of the text to English-language users only. With regard to the last point, it seems unlikely that the MicroGallery will be extended into other languages in the foreseeable future. The issue of color printing was raised when the system first opened and, at that time, the only printers available to color print at quality comparable to the image on screen would have been prohibitively expensive. However, as Ben Rubinstein stated in his presentation, there is now an archive of twenty-four-bit scans of the National Gallery's 2,200 paintings. He suggests that this information may be used in the future by setting up a dedicated station to print images on demand, perhaps through the gallery shop.

Although there is a slight variance between the MicroGallery in the Sainsbury Wing and the CD-ROM version, the differences are marginal and depend on the adaptation of the product to museum or schools and home use. The broad outlines for both are roughly the same. There are five major sections: "Artists' Lives: Lives of the artists and their paintings in the collection," "Historical Atlas: The collection organized by place and time," "Picture Types: The collection organized by types of work," "General Reference: An illustrated glossary for the collection," and, finally, "Guided Tours: Four narrated tours illustrated with highlights of the collection."

Each of the five sections can be cross-referenced; so, by taking Antoine Watteau as our example, one can access a short account of his thirty-seven years, consisting of five paragraphs of information with the option to further cross-reference the painters who influenced him and the painters he influenced. Here, one can go to Rubens and Teniers in the first instance and then see how his genre of the *fête galante* was taken up by Lancret and Pater. The *fête galante* and the French Academy are treated in additional pages; one can access the former, along with the Rococo period and the theme of Arcadia, in the general reference section, and then move on — maybe into a brief outline of Paris from 1700–1725. The choice is yours.

While watching the delicate pixels that have now reformed Watteau's "The Scale of Love," the most interesting option is to touch or click on the painting's classification as one of nine forms of "Everyday Life," that of "Making Music." This is part of the "picture types" index, and the MicroGallery's taxonomy of the representations of everyday life makes for an interesting browse. This section is split into nine "genres for living": making music, courting and brothels, drinking, leisure and games, education, work, soldiers, smoking and blowing bubbles, and domestic interiors. By now, a user of the publication, familiar with academic art-historical discourse, will realize that the information accessed is that of the canon — but given a social and historical twist that is more in keeping with Arnold Hauser than the breakthroughs made from the early 1970s onward. It is clear, though, that we are not dealing with the academy and, by the time you've left "Watteau" for "Everyday Life," the latter may turn out to be more interesting in terms of where you can go next. For instance, if you return to the Paris of 1700–1725, it may well be in the company of another painter — or you may be intrigued by another category altogether.

The key to the seduction of the MicroGallery lies in navigating the information and not in the information itself: the latter is always subordinated to the former. The seduction is the lure of the labyrinth and the aura of technology: one moves through it in the manner of a rhizome rather than in a linear sequence. The question that most reviews raise is, as Sharratt writes,

> What to do with it all? . . . The CD-ROM provides no overall argument, and its several layers of texts only suggest and enable connections, and point toward received views and established interpretations.

What we have, in the end, is a canonical hypertext or an armchair toy for the tired businessman: there is nothing to do but be absorbed in navigation. Judging by the entries in the comments book, users of the system appear to be continually absorbed, which suggests that the aura of pigment has given way to the aura of the pixel.

The process of navigating the MicroGallery reduces the paintings to empty tropes of themselves. This mode of electronic reproduction hermetically seals the images within a representation of the actual gallery. The MicroGallery offers the ontology of a "world" in which one does not have to spend time building a relationship with a mute object: paintings change over time as does the beholder — one will never come to the same picture twice. The essential estrangement of the pictorial world demands time to navigate and to see, and the illusion a painting produces, which *is* linked to its aura, cannot be reproduced: this is the liminal space the beholder enters to experience a picture.

It is to the credit of the MicroGallery design that one *does* become absorbed in the simulacrum through the power one cathects onto the technology, being drawn under its spell in a manner not unlike other CD-ROM games such as "Myst" and "Doom." The most interesting part of the MicroGallery lies, I think, in its animations: although designed to illustrate the representations of the paintings, they have the effect of off-setting the "classicist" graphics by bringing some subtle moments of comedy. For example, in an unwitting parody of Terry Gilliam's animation for Monty Python, you can watch the left-hand section of Poussin's "Bacchanalian Revel before a Term of Pan" rotate across the screen to its repainted mirror image in "The Adoration of the Golden Calf." The section is then rotated and reversed back before its term of Pan. It's an extremely funny moment and, ironically, it is in virtual keeping with the myth of Poussin painting from figurines on illuminated model stage-sets. One would have liked to have seen an animation of Anthony Blunt explaining and remarking upon this *détournement* of the master. It would also be a rare chance if one were allowed to open and close "The Donne Triptych" or Guisto de Menabuoi's "Coronation of the Virgin" — the MicroGallery can offer you that pleasure of technological beholding. It also allows you to press "with arch" or "no arch" on Uccello's "Rout of San Romano." In the end, I preferred the painting "with arch" and then decided to move to the animated reconstruction of the Camerino d'Alabastro, in Ferrara Castle, so I could see the Titians hung together. It was a beautiful moment: in the manner of Denis Diderot's salon criticisms of the 1750s, in which he applied the critical fiction of physically entering a depicted scene, I realized I was becoming lost in a Peter Greenaway movie, *Drowning by MicroGallery*. The only way out was to go through the simulated emplacement of Holbein's "The Ambassadors," watching the anamorphic skull rotate in a virtual half-life. I ended my twenty minutes at the terminal by watching the Muybridge animation of the human form in motion historically misplaced in the pages on Andrea Mantegna's "Introduction to the Cult of Cybele." I didn't have enough time to verify a visitor's complaint that the "Cubism" entry states that Picasso's "Les Demoiselles d'Avignon" is in the Metropolitan Museum rather than the Museum of Modern Art in New York City.

So what to do with it all? In reviews, the standard textual trope for dealing with the technological pleasure of the MicroGallery is that of Walter Benjamin's essay "The Work of Art in the Age of Mechanical Reproduction." It still offers a route into thinking through the relationship between the National Gallery's collection and its multimedia reproduction. Benjamin's insight that the aura of the painting is linked to its ritualistic function (magical, religious, or in the secularized form of beauty) allows one to realize that the MicroGallery is a form of ritual in and of itself — independent of the domain of real pigment.

As a commodity, the CD-ROM makes the perfect gift to round off a day in the gallery. Though, as a couple from New York proved while I was moving through the system, people will now shop for the CD-ROM museum before using the MicroGallery or seeing the collection itself. As with other shopping impulses, it is best to get the fetish just in case it is sold out while you are looking at the paintings. The iconographic program, which must begin and end with the consumption of souvenirs, has now found a temporary apotheosis in the technology that allows you to leave with a version of the museum itself. This is an anesthetic ecstasy of the Baudrillard-drome and an introduction to the cult of the pixel.

A few last words. It is difficult to write about the MicroGallery outside the format of a review. The system is encyclopedic and, within its "world," provides its own context and justification. One can write in a mode of Luddism, which will always have the potential to exhibit nostalgia for the paintings that have themselves been extracted from their original location and function. The MicroGallery only provides a further curatorial twist, as an arcadia of pigment is relocated into an arcade of pixels. The MicroGallery creates the illusion of a "world" through which one is absorbed, and it is this process that forms the frame of mind that lets one forget the gallery. As with McDonald's, though, you always know what you're going to get. The system empties the paintings of historical time, as is the nature of reproduction, and reframes them in the museum-on-a-disc. It is the medium that, in the end, is the message, and its aura is one of novelty for, as Adorno rightly stated, "The new wills non-identity, but, by willing, inevitably wills identity." This is the paradox of the MicroGallery. Its "world" induces a forgetting that can only be substantiated through the remembrance of the system's pretext of the artifacts themselves.

*Phil Tippett*
*interviewed by Iain A. Boal*

# IN THE TRACKS OF *JURASSIC PARK*

*How have the new digital technologies affected your work as an animator?*

There is a hysterical feeling that if you don't use digital technology, you cannot participate.

Experiments began years ago with off-the-shelf, inexpensive Macintoshes. Pictures were put together in ways that had not been possible before. In doing motion pictures, often an elaborate and costly device called an optical printer is used to composite images. Such a device employs multiple camera and projector heads that run specific color-matched pieces of film through finely machined gates with finite tolerances, so that the film doesn't wiggle or pop — the lensing technology is also amazingly refined. The operators of these optical printers are very specialized — considered to be black magicians of the trade — and there are only two or three people in the world who are adept at doing this.

Raw negative film is sent out to be processed and returned. There is always a lag between what you're shooting and what you're seeing. You may spend a day shooting and then most of the next day figuring out what you did the previous day. The computer allows all the images to be combined on one screen and viewed together. Representational photographs can then be made of the composited images.

Pretty much the same thing has happened in other areas of visual arts — in desktop publishing and illustration, for instance. The rise of the computer over the past few years has short-circuited a lot of careers — people have had to turn around very quickly and play to the demands of clients who want everything done on the computer. Recently, this has begun to change because a lot of computer work is starting to look the same.

What the computer has allowed us to do is to essentially put together a studio and integrate a lot of otherwise expensive studio processes such as optical printing, adding or mixing sound effects, cutting — these were entire departments at a studio, with huge crews of hundreds of people and heavy overhead. The hope is that this new technology will allow for smaller groups of people to work more independently, keep costs down, and make more diverse and interesting material.

*Is this driven by the big studios trying to cut their costs?*

Absolutely. It's difficult to characterize from my perspective — I view it as being hysterical. It is sort of an extension of business and technological communities supporting each other. A lot of it makes no sense on an intuitive level; there is a utopian, pie-in-the-sky aspect to all of this, that somehow this technology will make everything easier, better, and cheaper, and allow everyone involved to make more money. It will involve the entertainment market but will expand to all sorts of different "eco-niches" globally. Everyone is talking about a multibillion-dollar global marketplace. I see a stampede toward that, whatever that turns out to be.

*How will this affect the way films look, in general? Will they look better? Was* Jurassic Park *made better than it ever could have been using more conventional technology?*

For *Jurassic Park,* the technology certainly allowed for a different look — that look replaced "artifacts" that the viewing audience had found objectionable in the past. Using conventional technology to get the "slickness" required of a $60-million motion picture is just not feasible, and is becoming more and more difficult over the years.

*What do you mean by "conventional technology"?*

From the perspective of doing special visual effects for a so-called class A theatrical feature film, this involves a great deal of computerized robotics and an incredible amount of heavy industry — lots of machining work to build gantries and tracks — lots of robotics work. Repeatability: running the optical printers so they can do multiple passes on frames at varying exposure lengths — track cameras had to be able to repeat a move down to tens of thousandths of an inch, over and over. For performance animation, very elaborate and extremely cumbersome computerized puppets that allowed us to go in and refine performances over and over, rather than generating a performance under the constraints of time and budget, and hoping it would work — while working with the Lucases and Spielbergs. The technologies that we had been brought up with had been around since the inception of film, developed by George Méliès and Edison. We used those kinds of techniques exclusively up until about two years ago — exactly the same techniques: putting pieces of glass in front of the set, painting on the glass and shooting through it. We even used exactly the same cameras that were developed in the '20s and '30s.

What I did with my crew was to refine the innovations that people had made fifty years before. We used the same techniques but took them to a new level of sophistication. These techniques, used to enhance production values in the '50s and '60s for extremely low-budget movies, were used in the '70s and '80s for thrilling effects in big-budget Spielberg movies, where a great deal of attention had been put on special effects.

*What was driving the increased attention to special effects?*

Initially, it was the childhood dreams of a number of us who grew up in suburban America — we had grown up on the science-fiction films of the '50s and '60s, and now we were old enough to enter the workplace and carry on that fascination — that love. That's what Lucas, Spielberg, I, and many others were all about. There's a tremendous working relationship as well, with very little explicit communication — we all knew what the others were talking about: "Oh, this needs to look like *Forbidden Planet* but not like *The Island of Dr. Moreau. . . .*" Everything had to look better than whatever we had done the year before. We carried on like that for about fifteen years — each time trying to outdo ourselves and make more engaging performances than before.

The issue with creating characters that can exist in no other way than in theatrical feature films is about being as photo-representational as you can possibly be. That is a consequence of shooting all of your backgrounds with live actors who live in that movie space of twenty-four frames per second. The color, density, and type of film — all of that has a particular look. Everything has to be made photo-representational, so that you can't tell the difference between the photographed object and whatever is being pasted into it.

Conventional stop-motion photography involved models and a succession of still-motion frames — each frame had to be absolutely clean and clear, and had a sort of staccato effect — rather than the characteristic motion-blur of photographing a person or something in real time. A lot of time and effort has gone into trying to digitally re-create that motion-blur or continuity of action.

*Is this the same as trying to achieve "realism"?*

I prefer the term "photo-realism" — we're trying to match something, make it like something else, make it match an image in a photograph. Everything we do in the three-dimensional world — making models and photographing the models — has ultimately been to create a two-dimensional object — the movement that is projected on the screen.

Everything changed overnight with *Jurassic Park*. I like to use the example of the 1925 fantasy film *Lost World*, which was about an expedition to South America to bring back prehistoric animals. It was created using the stop-frame technology — the most advanced, most sophisticated use of the process at that time! If you read the reviews of *Lost World*, the critics back then were ecstatic — they couldn't believe it! They thought the filmmakers must have gone into the South American plateaus to shoot real dinosaurs! My kids watch the movie today and say "My God! — it looks like they were made of papier-mâché and clay!"

*So you have the feeling that the computer graphics images of today will look that archaic in the future?*

Certainly in the next ten years there will be something that replaces current technology — that looks "more real," in movie language. If you ever go into a meeting with the directors and producers and ask them what they want, they always say "I want it to be *more real!*"

For instance, when I was working on the models for *Jurassic Park*, I had a weekly video conference with Spielberg, who was off filming *Schindler's List*. Now, the video hook-up was only one-way: Spielberg could see me and the dinosaurs, but I could only hear him. Imagine the weirdness of the scene — here I am holding a model up to the lens of the video camera in Berkeley while Spielberg yells, "More real, Phil! More real!" from his end, in virtual Auschwitz!

*How is the digital technology going to change the motion-picture industry?*

I think the industry's main concern is with cutting costs, controlling the budget and the product. Historically, when the studios first developed out of the nickelodeons, they were complete bastions — each studio was its own fortress with its own group of actors, directors, cameras. By the '60s, the studios had become too top-heavy, television had taken the audience away. . . . People didn't start coming back into the theaters until the mid-1970s, really. Studios started to make more money, and turned into bureaucracies again. For a while, between the mid-1960s and mid-1970s, there was a tremendous opportunity for young directors — Lucas, Spielberg, Scorsese — today's main moneymakers. There was no real hierarchy in the studios — no one was in control.

Now that the motion-picture racket has become a $3- to $5-billion-a-year industry, the studios have come full circle, turning back into the movie bastions of the '30s and '40s. I believe their sights are set on the potential that the new technology will bring — being able to develop entire projects in-house, on the lot, and then being able to electronically distribute that material. Advertising and distribution add about $10 million to $20 million on top of a $20 million to $60 million picture. So studios are looking for the savings that they'll have with the fiber-optic superhighway. Just the mechanical distribution process is phenomenal — for a big hit, you have to distribute, say, 1,500 prints of a movie. Just the chemical processes are a big deal: running that much film through all of the processes and having each of those reels come out the same color and density is an extremely expensive technological nightmare. This will end, easily in the next ten to twenty years, maybe sooner. In fact, just last year, Sony piped *Dracula* into a theater over cable.

If the information superhighway gets up and running and movies-on-demand becomes popular — if everyone in America decides they want to watch first-run movies in the sanctity of their own home — that will set up a domino effect on the revenues for the theaters, video rental houses, and so on.

*Then, do you think that this quest for economy and control will change the way movies are made as well as distributed?*

Well, just try and imagine what will happen when technological giants like Microsoft and Silicon Graphics develop boxes that are complete, self-contained studios that will do everything — with digital input, processing, and all that — you can make a movie in a box and pipe that information into television set or theaters.

The entertainment racket itself is moving towards the idea of virtual reality. What does this mean for Disney and Universal when all you need is a headset and a closet to create an environment, a virtual community? Do they need to keep building these multibillion-dollar theme parks when all of this can be done in a little black box?

I can't imagine something that would catch on as much as theatrical feature films did, or television. Television is just more accessible — more on the level of the human scale: it's like sitting down together in a group in front of the sorcerer and listening to stories. . . . We do have some background in that. Theatrical feature films have always been churchlike to me, with rows of people going into a darkened environment, willing to suspend their disbelief as the big organ blasts out all this stuff!

Ultimately, on the media end, everything has to revolve around stories or an adrenaline rush — some kind of a ride. Is a computer game a story or a thrill? One thing these games represent is money — everyone talks in billions.

*Perhaps there is a misconception about what virtual reality is capable of in relation to how people — is it really that fascinating?*

I don't know! I don't think anybody knows anything! There *is* a lot of speculating going on — a lot of people would like to be preemptive and say "This is the marketplace of the future." But are we even going to care about the whole digital realm? *I* don't care about the virtual world — can't stand to look at a television screen for very long. The members of my crew go crazy — they can't stand to sit in front of the television for that long. For the younger ones it's not a problem ... they're absolutely at one with it. The thought of negotiating in a virtual world — what's the point, the purpose? I just don't understand. Maybe it's a generational thing.

There are so many people that are looking in this mad rush for something with some certainty ... unless civilization crashes and burns, digital technology is going to take off like an industrial revolution, and have a profound effect on the whole world. But the things that are interesting to me are the same things that were interesting to the guys who painted by torchlight on the walls of the caves in Lascaux. Performance, drama, picture-making, making objects — these are the things that are compelling to me — not a virtual reality I can't possibly imagine.

*Are the old movie-making skills being replaced by the new technologies?*

The *tools* are being replaced. This is another part of the digital hysteria — because these tools are so unwieldy, so counterintuitive and difficult to use. . . . There are two tendencies in the industry right now: one tendency is to hire as many qualified computer graphics technicians as you can, and teach them how to be picture makers or actors or whatever. I don't think this is going to be successful. On the other hand, the emphasis in my studio is to work with a minimal number of computer people, and train people with more traditional skills to work on the computer.

*If the tools are so hard to use and so expensive, why has there been this tremendous investment in digital technology?*

Digital technology gives a great deal more participation to the business community, the managers. Everything in the studios right now is very heavily managed — there's not a filmmaker among them: they're all MBAs, lawyers, and agents. That's who runs the motion-picture racket. There are no directors, producers, or actors in management.

I believe they find the computer realm extremely attractive because it looks like what they're used to! It looks like accountants working together in these tiny controlled environments — it's clean, fast, and logical! It's not that down-and-dirty carnival atmosphere. It's clean rooms, clean clothes, well-shaven, well-kept people. Sitting in front of a computer monitor gives them that "hands-on" feeling. And their job is to make movies pay — they've gone to school to make this a career. They can look at "the bottom line" and decide what they do and do not want in a picture.

And digital technology has allowed filmmaking to get sloppier: if there's a problem with a setup, you can fix it digitally. In *Jurassic Park* a scene accidentally included one of the grips — the people who move the lights around — standing in the shot, smoking a cigarette. This kind of mistake can be digitally removed.

*Was the film world all that different when you started out?*

For everyone I know, filmmaking was a calling — you didn't think at all about making money at it. It was more like "I've *got* to be in the circus — I *will* be in the circus!" It was that kind of mentality, up until about five years ago. It had more of a carnival atmosphere and a sense of community. There used to be a great deal more attention to craft. . . .

There *are* still people here who are craftsmen. What we're trying to do here is train these people on the computers, and make our workshop a place where a computer can be a powerful and important tool that we use like lathes and mills. It *is* very exciting to be able to make virtual clay — virtually manipulating it!

*There must be a tremendous capital investment tied up in all these machines. How is your business able to keep up with such rapidly changing technology?*

We're a beta test site for a number of computer graphics companies. We get to tell these companies what we'd like to see in these tools. For small mom-and-pop companies like us, a lot of workstations alone is not enough. If you want to have the tools you need, you have to have support from the companies that build the tools.

We have a reputation, primarily because we do things the old way — we're using these tools the way God intended them! — in our work as artists, craftsmen, and performers. And it's yielding a very different kind of product than what you see anywhere else. We help the computer companies with their advertising, and they help us with their software.

*Is using a computer as a tool much the same as using a pencil or a hammer?*

Any craftsman knows that to use tools to craft an object, you have to know your tools and their limits. The computer world is an area where people have no clue as to what can and can't be done with the tools. If you don't have an idea of what you want to do, and what the limitations of the tools are, then all of this stuff is going to come out looking pretty much the same. And from my perspective, 99 percent of what I've seen created on computers looks the same — a very small amount of it is interesting, just in terms of conventional picture-making or performance or drama. It's like airbrush art — you look through books of the incredible stuff people can do with airbrush, but somehow it all looks like the same picture.

*Don't you find it very strange that the rise of this computer hysteria happened to coincide with the collapse of the Cold War?*

Sure. It's as if someone said, "What are we going to do now?" Lawrence Livermore Labs is moving out of the bomb business — now they're working on compression schemes to squeeze information smaller and smaller and smaller. From my perspective, the entertainment racket has just taken off — gone from a carnival mentality to absolutely voracious multinational companies banging each other on the heads. And here we are, this little company sitting here watching these huge rockets flying overhead. . . .

Another thing to think about — this makes my brain hurt — is the $40 thousand or $50 thousand for each of these workstations. You know I'm never going to make any money doing that! It's not an investment — I'm not investing in a piece of equipment that I'm going to be able to sell! This stuff is worthless in a matter of months! There's a continuous drive to put out more powerful new computers. My initiation fee to get involved with this digital realm has been really high. It's like putting coins in a pay phone: if you want to talk, say goodbye to your money. And the frustrating thing for me — having come from being a model maker — all my life I've made things with my hands.

But this technology is inaccessible to me — I don't have time to learn how to work these things, so I don't have any hands-on experience anymore! Do I want it? Who wants to sit down in front of the television set and switch channels all day! I'd like to *make* something — but it's not economically feasible here. The people who work on the machines understand them much, much better than I.

*Is this new technology going to be somehow more democratic, in the sense that people could be making their own movies and animation?*

It hasn't really happened that way, because you still need a publisher or distributor. And everything's extremely expensive. Democratic? Holy smokes! You'd have to be a multimillionaire to come close to using this junk on any meaningful scale!

*James Brook*

# READING AND RIDING WITH BORGES

Jorge Luis Borges hated his job in the municipal library in Buenos Aires — his first steady job, secured in early middle age. It's said that he would cry from boredom and frustration on his long commute home on the tram; Borges was a member of the sinking middle class, and he had imagined better things for himself. His menial work seemed a tedious detour from his fate as a serious writer with a place in society. Yet the empty trip — mere duration over desert space, the definition of any commute — gave him many hours to read Dante and Kafka, authors of nightmares of labyrinth and punishment. "Midway in the course of our life's journey / I found myself in a dark wood, / for the straight way was lost." Lost perhaps in an evil reverie of a penal colony where "a remarkable piece of apparatus" inscribed the prisoner's "sentence" on his body.

A dull job in the stacks gave Borges the freedom not only to dream but to inscribe nightmares, nightmares that accurately figured the computing machines that were then being developed for a world hurrying to war — and prefigured those enthusiasts of informationism, certain postmodern critics and writers. In the dark visions of "The Total Library," "The Library of Babel," and "The Garden of Forking Paths," Borges conveyed his premonitions of the world the cheerful industrialists of the communication apparatus are preparing for us.

James Brook is said to have cried at least once on the way to his dull job in Silicon Valley. He often forgot — vertigo and panic! — whether he was coming or going on the slow train that shuttled between San Francisco and the Peninsula. Automobile traffic crawled down the freeway that paralleled the tracks, the solitary drivers gazing at the hyperreal landscape through TV windshields, their "sound systems" relaying the voice of a fascist talk-show host or the synthesized notes of a pop tune or the cultural tones of a recorded book declaiming or blaring or declaiming. Brook didn't cry because the job was a detour from his destiny — rather, he felt that the job *was* his destiny and that the wired world he was inadvertently helping create was the planet's destiny: a place that excluded the experience of the time of the journey, that excluded the time of experience; a place where all traces of nature had been transformed into industrial parks, all knowledge into bits of information, all adventure into a closed game. Yet in a world that never seems to shut up for a minute, the daily train trip gave him time to read Borges, to write this essay, and to tell a friend about the last stanza of Tom Raworth's poem, *The Vein* (1991):

> hardly dimmed the harsh light
> he sometimes pulled at his hair
> obsessed with finding the beautiful
> curtain allowing him entry
> never able to follow
> the middle of night
> downwards to find runway
> with deep sides
> writhing under his fingers
> personalities full of energy
> order a series
> of the same programme
> cool for film
> using this knowledge
> machines talk to themselves
> maintain a very persistent
> buzzing as the signal
> ends in a dramatic freeze

          close to the border
          on a street with a few orange trees

A hell of recent devising, where "machines talk to themselves" and "the signal ends in a dramatic freeze" amid the last few orange groves of the once fertile land: this was Brook's stop. Here is where he would attend to his small part of the not-quite-thought-out jumble of plans to speed up the global flow of information that was entertaining to some, useful to others, and speculatively profitable to unnamed transnational corporations.

·

"The Library of Babel" and its precursor essay, "The Total Library" describe the labyrinth of books, the combinatory of the alphabet (a technology whose effects still stagger us), and the hell of information considered in its architectural aspect. Half a century before the proposals for the grand antipublic works of converting the Bibliothèque de France and the Library of Congress to on-line information systems, Borges matched the vision of total information — most of it necessarily meaningless — with an image of mad organization, of lunatic encyclopedism linked to the mechanized generation of endless texts.

     "The caprice or fancy or utopia of the Total Library contains certain traits that could be confused with virtues," Borges warns as he traces the history of the fantasy of total information derived from the infinite combination of the letters of the alphabet (1981). Borges follows the thread from Aristotle, who "reasons that a tragedy is made up of the same elements as a comedy — that is, the twenty-six letters of the alphabet," to Cicero, who parodies the theory of "the fortuitous collision" of atoms producing this "elaborate and beautiful world," to Thomas Huxley's invention of the trope that "a half-dozen monkeys, supplied with typewriters, would produce in a few eternities all the books in the British Museum." (Borges notes that the scheme really requires only a single monkey and one eternity.) Not only would "all the books in the British Museum" eventually be reproduced by this method, but so would every possible book that is *not* in the British Museum.

     In "The Total Library," it is Cicero who first makes the connection between the combinatory powers of the alphabet — so different from those of, say, Chinese characters — and the powers of money when he declares that "if a countless number of copies of the one-and-twenty letters of the alphabet, made of gold" were "shaken out on the ground," he doubts "whether chance could possibly succeed in producing even a single verse!"

In Cicero's off-hand disparagement of the combinatorial resources of quantified knowledge and wealth we find the possibility of conceiving of the project of information theory and its vertiginous reduction of communication to *senders* and *receivers* exchanging encoded *messages* — with the meter running. This recombination of elements relies on a prior analytical move, the radical destruction of meaning as inherently expressive *words* are broken down into inarticulate *letters* — which will later serve as models for the *codes* of telegraphy, telephony, and computing.

That every code is so much less than any human language is a proposition enacted by "Téléphone," a set of two plaques — one "positive" and the other "negative" (in at least the photographic sense) — produced by Marcel Broodthaers in 1968:

*Je suis fait pour enregistrer les signaux.*
*Je suis un signal. Je Je Je Je Je Je Je Je*
*Objet Métal Esprit Objet Métal Esprit*

*Je suis fait pour enregistrer les signaux.*
*Je suis un signal. Je Je Je Je Je Je Je Je*
*Objet Métal Esprit Objet Métal Esprit*

☎   ☎   ☎   ☎   ☎   ☎   ☎   ☎

*Objet Métal Esprit Objet Métal Esprit*
*Objet Métal Esprit Objet Métal Esprit*
*Objet Métal Esprit Objet Métal Esprit*

[I am made to record signals.
I am a signal. I I I I I I I I
Object Metal Mind Object Metal Mind
    *or*

Metal Object Spirit Metal Object Spirit
*and so on*] (Broodthaers 1988)

Critically mimetic of the cybernetic model of communication (repopularized in every decade since the 1950s), which reduces human communication to an exchange of signals and human language to machine-manipulable code, "Téléphone" makes explicit the usually casual identification between the human and the mechanical, and the ambivalent nature of this identification. The repetitions of words and phrases imitate the "call and response" of machine communications, which fight entropy — noise on the line. The mental becomes metal, a technically manufactured object, consciousness re-formed in the image of the transmitting and receiving apparatus.

As can be seen by looking back from "Téléphone," Cicero's golden letters mark the parallelism of two circuits of exchange: that of information and that of money. The metaphor of "golden letters" is not accidental: over the centuries these circuits — of quantitative abstraction of information and quantitative abstraction of exchange-value — have been united in and by the commodity form, just as Marx depicted it in the first volume of *Capital*. Which see.

Concluding the theoretical survey of "The Total Library," Borges informs the reader that he's described "a minor horror: the vast, contradictory library, whose vertical deserts of books run the incessant risk of metamorphosis, which affirm everything, deny everything, and confuse everything — like a raving god." That raving god was about to be "incarnated" in the computer, which was first designed and constructed as a processor of numbers but conceived by Alan Turing as a manipulator of symbols — the alphabet included.

It would take some time for a writer like Italo Calvino to be intrigued by the combinatorial capabilities of the computer, but in a 1967 lecture, "Cybernetics and Ghosts," Calvino (1986) — in step with his colleagues in Oulipo (Ouvroir pour littérature potentielle [Workshop for Potential Literature]), including cofounder Raymond Queneau — mixed then-fashionable structuralism with that year's fascination with electronic circuits. Projecting the origin of story telling back into a mythical past, Calvino evokes language itself as a kind of machine: "The story teller began to put forth words . . . to test the extent to which words could fit with one another, could give birth to one another, in order to extract an explanation of the world from the thread of every possible spoken narrative, and from the arabesque that nouns and verbs, subjects and predicates performed as they unfolded from one another." In Calvino's overly *linguistic* view of literary creation "writing is purely and simply a process of combination among given elements" and "writers . . . are already writing machines." The next step was to transfer the labor of creating literature to suitably programmed machines. The "decisive moment of literary life will be that of reading" the products of "electronic brains." In a complementary move, Calvino reflects (passively!) on the brain's electronics, "the transistors with which our skulls are crammed."

Nowhere in their decades-long investigations into combinatory literature did the Oulipians (self-described as "rats who must build the labyrinth from which they propose to escape") ever remark on the absurdity of trivializing literature by rendering it so completely subjective — subjective thanks to its very objectification in a machine-produced object. Now, the programmers of combinatorial poems, romantic stories, and hypertext webs glory in "the death of the author," whose demise, in this uninspiring rehearsal of the Christian drama, mysteriously "empowers the reader," who pays to consume what the more honest industrialists call "the product."

•

In *my* reading of the parable of "The Library of Babel," Borges sketches the imaginary physical architecture of total, encyclopedic information, where "The universe (which others call the Library) is composed of an indefinite and perhaps infinite number of hexagonal galleries, with vast air shafts between, surrounded by low railings" (1964b). No cybernetician could challenge the exactness of the technical description of the self-regulating, ahistorical archive that has no need of the dying breed of ineffectual "librarians" who wander through its endless galleries that are packed with volumes of the nonsense of random strings of alphabetic characters. The "Library is total and . . . its shelves register all the possible combinations of the twenty-odd orthographical symbols. . . ."

And no cybernetician could see this refurbished circle of hell as anything but a catalog of the virtues of paradise, despite the story's insistent evocations of death and despair in the face of the inexhaustible archives of information and despite "the formless and chaotic nature of almost all the books." (The narrator mentions that one book "is a mere labyrinth of letters, but the next-to-last page says *Oh time the pyramids*" — a glimmer of apparent meaning that could only increase the misery of the reader.)

Like scholars in the nonfictional world, the librarians' physical needs are minimally attended to: they are provided "insufficient, incessant" lighting in which to conduct their research and "To the left and right of the hallway there are two very small closets. In the first, one may sleep standing up; in the other, satisfy one's fecal necessities." Add a vending machine to the picture and advance fifty years to the preferred working conditions of computer programmers, who, like the adepts who "hide in the latrines with some metal disks in a forbidden dice cup," "feebly mimic the divine disorder" in celebration of compulsive rituals.

When Borges's narrator notes that "The certitude that everything has been written negates us or turns us into phantoms," is he commenting on a text that already contains the stifled laments and small pleasures of the postmoderns? In any case, in this plotless parable of the extreme spatialization of information, there is no room for consequential human intervention — it doesn't matter whether the books are read, understood, or even destroyed — and no possibility of human history. The librarians of Babel haunt the stacks of useless and alien information — information that they cannot turn into human knowledge. This uselessness is expressed in the narrator's admission "that this vast Library is useless: rigorously speaking, *a single volume* would be sufficient. . . ." (Replace "volume" with "database" and you have *one* goal of the visionaries of information systems. . . .)

A librarian throws himself into the air shaft that runs past his hexagon: death in the Library emanates from the pervasive estrangement of thought from the body and its desires, now just so much refuse.

•

The algorithm is the logical processor of the data that is input to the computing machine — and the machine is, as Alan Turing discovered, first of all a mental construct and only later — a very few years later, as it turned out — a mass of complicated electronic circuitry. Turing's "universal machine," the machine that could emulate an undetermined number of other machines, receives symbolic information, transforms it according to instructions, and produces some representation of that transformation. Indifferent to the meaning and nature of the symbolic data, the machine requires only that the data be in machine-acceptable form. The machine doesn't understand, it operates.

If one can read "The Library of Babel" as a parable of what used to be called the computer's "store" and which is now confusedly called its "memory," then one can read "The Garden of Forking Paths" (1964a) as a parable of the algorithm, the game of data in, data transformed, and data out, which Borges presents as the fragment of a thriller.

In a story behind "page 22 of Liddell Hart's *History of World War I*," a Chinese national in England to spy for the German Reich flees from an Irish counterespionage agent. His time is running out, but he must somehow send a signal to his handler in Germany to pass on the secret location of a British artillery park in France. "Thus I proceeded as my eyes of a man already dead registered the elapsing of that day. . . ."

The man already dead plays a deadly game: a step ahead of his pursuer, the spy, Yu Tsun, jumps on a train to the country home of Dr. Stephen Albert, a man steeped in the language, literature, and philosophy of China. Albert mistakes the spy for someone from the Chinese embassy and welcomes him into his home; in the course of their conversation, Albert reveals that he's discovered the secret of the last works of the spy's illustrious ancestor, Ts'ui Pên, who, at his death, left behind an incomprehensible "novel" whose episodes did not respect time, cause, or effect, and the rumor of a labyrinth — one that may or may not have been built though it had certainly not been found.

It turns out that Ts'ui Pên had collapsed the novel into the space of the labyrinth and submitted the labyrinth to the time of "both-and" decision making rather than "either-or" branching: the labyrinth was the novel.

> Almost instantly, I understood: "the garden of forking paths" was the chaotic novel . . . forking in time, not in space. . . . In all fictional works, each time a man is confronted with several alternatives, he chooses one and eliminates the others; in the fiction of

Ts'ui Pên, he chooses — simultaneously — all of them. *He creates* diverse futures . . . which themselves also proliferate and fork. . . . He believed in an infinite series of times. . . . This network of times . . . embraces *all* possibilities of time.

This labyrinth of time presents the spy with an image of refuge from the inexorable pursuit that necessarily spells his death. The labyrinth in which he is caught is a story with only one possible outcome. Its time moves in just one direction, and it's a time reduced to the mechanics of clocks and train schedules — empty time indifferent to either human desire or natural cycles. Like any computer program — or any computer game — the story is relentlessly plot driven, submitted to binary "either-or" branching. Only fiction — the imagination working through and against time — can brake the linear career of the plot.

But the spy must complete his mission. As his pursuer approaches the house, Yu Tsun shoots Albert in the back: "Albert fell uncomplainingly, immediately. I swear his death was instantaneous — a lightning-stroke." The spy's chief knew that his "problem was to indicate (through the noise of war) the city called Albert," which was accomplished by killing "a man of that name" — this name sure to be printed in the English newspapers.

The machine has carried out its instructions, processing the data and emitting the required signal. Space has been annulled, the life's work of the sinologist has been reduced to a secret name, a code to be transmitted across the sea so that the war machine might continue its ruthless milling of men and cities. The game can result only in the multiplication of death.

Recently, a professor of English become "hypertextualist fiction" writer turned his gaze, hand, and keyboard on "The Garden of Forking Paths." Stuart Moulthrop (1991) "found signs that the world that lies before us may be significantly different from the paradise of the printed book." Significantly different are the hellish punishments Moulthrop inflicts on the Borges story and its readers.

Moulthrop is determined to go beyond the merely conceptual deconstruction of fiction: he wants not just to analyze how a "text" works, he want to reduce it to a kind of clockwork. Borges, with his stories about stories, is neither postmodern nor industrial enough. His efforts are "restricted both by the immutability of their medium (the printed page) and by social practices (authoritative texts, the law of copyright) to a single set of discursive practices." What do these insults to the writer mean? That Borges's stories resist final critical appropriation, that they keep the critic in a subordinate (if better paid) position — and that the physical, legal, and commodity form of the book are now viewed as constraints on the marketability of "text," the literary world's equivalent of that quantity known as "information."

Exacting his critic's revenge, Moulthrop translates "The Garden of Forking Paths" into an electronic hypertext — a sort of database of textual fragments mechanically linked for the reader's navigation through and rearrangement of them. Thus "Yu Tsun's vision of alternate selves would be no illusion. . . . [T]he reader could select a different way through the garden of forking paths" — the additional paths authored and scripted by Moulthrop himself, who supplies the "missing" scenes and developments!

All that in the Borges story unfolded in the time of reading is now spatialized, *mapped.* Moulthrop asserts the superiority of his deconstructed-and-reconstructed version: "The electronic 'forking paths' empowers its readers in ways that Borges can invoke only by hypothesis. It gives readers an actual choice of procedures and outcomes," like any game in an electronic arcade. This high-tech emulation of the combined attractions of *Reader's Digest* and the pinball machine enshrines Moulthrop's irremediable misunderstanding of the Borges story, which is based on maintaining the tension between the ever-tightening requirements of the plot and the increasingly remote refuge of the fictional labyrinth.

Scholar that he is — and alien to the spirit of the contemplative sinologist — Moulthrop has found a way to "go beyond" the book and the card file that the scholar feeds to the book. He recreates the book in another medium entirely, remaking it into mere support for the scholarly apparatus so well emulated by the universal machine. The machine has its instructions and so does the reader.

·

*Why combinatory texts are not Chinese poetry:* In classical Chinese poetry what is emphasized is the field of relationships between characters, irreducible characters infinitely more expressive than the isolated letters of the alphabet or the "textual units" of postmodern Newspeak.

.

Every industrial era — whether its industrialism is based on steam or information — has its nightmare and its dream. Borges described versions of hell, and some mistook these hells for heavens; but there remain other ways of imagining the labyrinth and its solution. In André Breton's "Introduction to the Discourse on the Paucity of Reality," written in 1924, "telephony," "adventure," and "imagination" are not immediately suspect or contradictory concepts:

> "Wireless," a term . . . whose success has been far too swift for it not to have carried with it a good many of our era's dreams and for it not to have provided me with one of those rare and specifically modern measures of our mind. It is faint gauges of this sort that occasionally give me the illusion that I am embarked on some great adventure, that I somewhat resemble a seeker of gold: the gold I seek is in the air. What do these words I had chosen indeed evoke? Hints of sandy coasts, a few daddy longlegs entangled in the hollow of a willow — of a willow or of the sky, which is no doubt no more than an extended antenna, then some islands, nothing but islands . . . Crete, where I must be Theseus, but Theseus forever trapped in his labyrinth of crystal.
>
> Wireless telegraphy, wireless telephony, wireless imagination, as it has been called. An easy induction to make. . . .
> (Breton 1994)

Not that the book is an eternal form or one suitable for all uses. Not that there aren't better methods of recording and communicating our thoughts. Life, poetry, and knowledge predate the book and, if the human race isn't completely out of luck, they will survive the book's demise and possible multiple transformations. Hasn't the book of writing already migrated not once but several times? Hasn't the book seen its contents spilled across the stage, the film, and the audiotape? Hasn't it been invaded by the photograph and the painting? Hasn't it been used to build the archive and the database? Hasn't it been sacrificed to the conversation between friends and the sweet obscenities lovers whisper in the small hours, with the radio on?

Yes, there is a beyond to the book — there always has been: whenever there has been human communication.

But the box or network powered by electrons and algorithms with all its furious button-pushing — all its performance of tawdry wonders — militates against one of the finest potions we have: reading as reverie, where the sovereign image is one origin of freedom. The adequate use of the book, once relieved of the burden of logging all the shit that might just as well be sent to the cathode-ray tube, is key to escape from the pressures of economic and productive life.

While recycled and resold information races by in the unseen fiber-optic cable, Brook reads and turns the page. He curses the laws of exchange that send him off to work. He makes a note in the margin: a few words in an indecipherable hand that rhyme the swaying train. He surrenders to the irruption of images, to the *unpredictable poetic,* which, despite the din, is only echoed by the *unpredictable mechanic.*

*References*

Borges, Jorge Luis. 1964a. "The Garden of Forking Paths." In *Labyrinths.* New York: New Directions.

———. 1964b. "The Library of Babel." In *Labyrinths.* New York: New Directions.

———. 1981. "The Total Library." In *Borges: A Reader.* Edited by Emir Rodriguez Monegal and Alastair Reid. New York: E. P. Dutton.

Breton, André. 1994. "Introduction to the Discourse on the Paucity of Reality." Translated by Richard Sieburth and Jennifer Gordon. *October,* 69.

Broodthaers, Marcel. 1988. "Téléphone." In *Broodthaers: Writings, Interviews, Photographs.* Edited by Benjamin H. D. Buchloh. Cambridge, Mass.: MIT Press.

Calvino, Italo. 1986. "Cybernetics and Ghosts." In *The Uses of Literature.* Translated by Patrick Creagh. New York: Harcourt Brace Jovanovich.

Moulthrop, Stuart. 1991. "Reading from the Map: Metonymy and Metaphor in the Fiction of 'Forking Paths.'" In *Hypermedia and Literary Studies.* Edited by Paul Delany and George P. Landow. Cambridge, Mass.: MIT Press.

Raworth, Tom. 1991. *The Vein.* Great Barrington, Mass.: The Figures.

# CONTRIBUTORS

Howard Besser is Visiting Associate Professor of Information and Library Studies at the University of Michigan, where he teaches courses on the impact of multimedia and networks; he also teaches at the University of California, Berkeley. His articles have apppeared in the *Journal of the American Society for Information Science, Library Trends, Visual Resources,* and *Museum Studies Journal,* and he has contributed chapters to books on multimedia, cataloging, educational technology, and organizing moving image materials.

Iain A. Boal is an Irish social historian of science and technics. He teaches at the University of California, Berkeley, and is now at work on a book and film on charisma and healing in the revolutionary decades of seventeenth-century Ireland and England.

James Brook is a poet and translator who lives in San Francisco and labors in Silicon Valley. His poems and essays have appeared in *Exquisite Corpse, City Lights Review,* and elsewhere; his translations include *Resistance* by Victor Serge, *The Lives of the Gods* by Alberto Savinio, *Panegyric* by Guy Debord, and, with Sasha Vlad, *Zenobia* by Gellu Naum (Northwestern University Press 1995).

Chris Carlsson cofounded *Processed World* magazine in 1981, where a version of "The Shape of Truth to Come: New Media and Knowledge" first appeared. He edited and designed the *Bad Attitude* anthology, and is a graphic artist by trade. He divides his unpaid time between a grandiose experiment in interactive multimedia, exploration of the bicycle as an antispectacular device, and family and friends.

Jesse Drew has exchanged soldering iron, multimeter, and oscilloscope for camcorder, Macintosh, and modem. He produces independent documentaries, writes about technology and the public interest, and is currently a doctoral student at the University of Texas at Austin.

Oscar H. Gandy Jr. is Professor of Communication at the Annenberg School at the University of Pennsylvania. He is the author of *The Panoptic Sort* (Westview 1993), *Beyond Agenda Setting,* and numerous articles and chapters on communication, information, and public policy. He recently served as a fellow at the Freedom Forum Media Studies Center at Columbia University where he investigated the role of the press in the communication of racially comparative risk.

**Daniel Harris** is writing a book on gay culture. He has published essays in *Salmagundi* (where "The Aesthetic of the Computer" first appeared) and *Harper's*.

**R. Dennis Hayes** writes about technology and politics. He is the author of *Behind the Silicon Curtain: The Seductions of Work in a Lonely Era* and is writing a book on the shortcomings of computers.

**Doug Henwood** is the editor of *Left Business Observer*, a newsletter on economics and politics; a contributing editor of *The Nation;* and author of *The State of the U.S.A. Atlas* (Simon & Schuster/Touchstone 1994) and *Wall Street* (Verso 1995).

**George Lakoff** is a linguist involved with developing new directions in the study of language and mind. He is the author of *Women, Fire, and Dangerous Things: What Categories Reveal About the Mind* and, with Mark Johnson, *Metaphors We Live By*. Their forthcoming book is a manifesto of "experientialist" philosophy, in which they elaborate their ideas on the embodiment of mind.

**Les Levidow** is a research fellow in the Technology Faculty at the Open University. He is the coeditor of several books, including *Anti-Racist Science Teaching* and, with Kevin Robins, *Cyborg Worlds*. He is also managing editor of *Science as Culture*.

**Marina McDougall** is a media arts curator at the Exploratorium, a museum of art and science in San Francisco. Her film essay, *If You Lived Here, You'd Be Home by Now*, depicts the sense of placelessness induced by American architectural, automotive, and electronic realms.

**Laura Miller** is a journalist and critic who has contributed to the *San Francisco Examiner, Harper's Bazaar,* and *GirlJock*. She is a contributing editor for *San Francisco Weekly* and a worker-owner of Good Vibrations, a sex-toy store.

Monty Neill is an education activist and a member of Midnight Notes. "Computers, Thinking, and Schools in the 'New World Economic Order'" is adapted from a longer piece on education to appear in a forthcoming Midnight Notes publication. Neill and Midnight Notes can be reached at Box 204, Jamaica Plain, Massachusetts 02130.

Chris Riding is an artist and theorist. Dividing his time between San Francisco, California, and Leeds, England, he is currently teaching in the Historical and Theoretical Studies Department at Derby University, England.

Kevin Robins works at the Centre for Urban and Regional Development Studies, University of Newcastle upon Tyne. He is the author, with Frank Webster, of *Information Technology: A Luddite Analysis* and *The Technological Fix.*

Herbert I. Schiller is the author of *Culture, Inc.: The Corporate Takeover of Public Expression, The Mind Managers, Mass Communications and American Empire,* and many other books. He is currently Visiting Professor at New York University and Professor of Communication Emeritus at the University of California, San Diego. A version of "The Global Information Highway: Project for an Ungovernable World" will appear in *Farewell to the Common Good: Making Information "Haves" and "Have-Nots"* (Routledge 1995).

Richard E. Sclove is the author of a forthcoming book on technology and democracy (Guilford Press). He also directs the Public Interest Technology Policy Project, a collaborative effort of the Loka Institute and the Institute for Policy Studies. He can be reached at the Loka Institute, P.O. Box 355, Amherst, Massachusetts 01004-0355; e-mail: resclove@amherst.edu.

John Simmons is the author of *The Scientific 100* (forthcoming); his novels include *The Sharing* and *Midnight Walking.* He lives in New York and Paris.

Rebecca Solnit is an activist and author of *Savage Dreams: A Journey into the Hidden Wars of the American West* (Sierra Club 1994) and *Secret Exhibition: Six California Artists of the Cold War Era,* as well as many essays on landscape and public space in anthologies such as *Visions of*

*America: Landscape as Metaphor at the End of the Century, Compassion and Protest: Recent Social and Political Art,* and *War After War.*

**Phil Tippett** is an animator whose mentor was Tex Avery. Phil's creatures — the space bestiary in the *Star Wars* trilogy, "Vermathrax Pejorative" in *Dragonslayer,* the tyrannosaurs and velociraptors in *Jurassic Park*— have won him two Oscars and a secure place in people's nightmares.

**Ellen Ullman** has worked in software development for fifteen years as a programmer, engineer, and consultant. She is also a consulting editor for *Byte* magazine and a contributing author for *Red Herring,* a technology-investment monthly.